Communications and Control Engineering

For further volumes:
www.springer.com/series/61

Series Editors
A. Isidori • J.H. van Schuppen • E.D. Sontag • M. Thoma • M. Krstic

Published titles include:

Stability and Stabilization of Infinite Dimensional Systems with Applications
Zheng-Hua Luo, Bao-Zhu Guo and Omer Morgul

Nonsmooth Mechanics (Second edition)
Bernard Brogliato

Nonlinear Control Systems II
Alberto Isidori

L_2-Gain and Passivity Techniques in Nonlinear Control
Arjan van der Schaft

Control of Linear Systems with Regulation and Input Constraints
Ali Saberi, Anton A. Stoorvogel and Peddapullaiah Sannuti

Robust and H_∞ Control
Ben M. Chen

Computer Controlled Systems
Efim N. Rosenwasser and Bernhard P. Lampe

Control of Complex and Uncertain Systems
Stanislav V. Emelyanov and Sergey K. Korovin

Robust Control Design Using H_∞ Methods
Ian R. Petersen, Valery A. Ugrinovski and Andrey V. Savkin

Model Reduction for Control System Design
Goro Obinata and Brian D.O. Anderson

Control Theory for Linear Systems
Harry L. Trentelman, Anton Stoorvogel and Malo Hautus

Functional Adaptive Control
Simon G. Fabri and Visakan Kadirkamanathan

Positive 1D and 2D Systems
Tadeusz Kaczorek

Identification and Control Using Volterra Models
Francis J. Doyle III, Ronald K. Pearson and Babatunde A. Ogunnaike

Non-linear Control for Underactuated Mechanical Systems
Isabelle Fantoni and Rogelio Lozano

Robust Control (Second edition)
Jürgen Ackermann

Flow Control by Feedback
Ole Morten Aamo and Miroslav Krstic

Learning and Generalization (Second edition)
Mathukumalli Vidyasagar

Constrained Control and Estimation
Graham C. Goodwin, Maria M. Seron and José A. De Doná

Randomized Algorithms for Analysis and Control of Uncertain Systems
Roberto Tempo, Giuseppe Calafiore and Fabrizio Dabbene

Switched Linear Systems
Zhendong Sun and Shuzhi S. Ge

Subspace Methods for System Identification
Tohru Katayama

Digital Control Systems
Ioan D. Landau and Gianluca Zito

Multivariable Computer-controlled Systems
Efim N. Rosenwasser and Bernhard P. Lampe

Dissipative Systems Analysis and Control (Second edition)
Bernard Brogliato, Rogelio Lozano, Bernhard Maschke and Olav Egeland

Algebraic Methods for Nonlinear Control Systems
Giuseppe Conte, Claude H. Moog and Anna M. Perdon

Polynomial and Rational Matrices
Tadeusz Kaczorek

Simulation-based Algorithms for Markov Decision Processes
Hyeong Soo Chang, Michael C. Fu, Jiaqiao Hu and Steven I. Marcus

Iterative Learning Control
Hyo-Sung Ahn, Kevin L. Moore and YangQuan Chen

Distributed Consensus in Multi-vehicle Cooperative Control
Wei Ren and Randal W. Beard

Control of Singular Systems with Random Abrupt Changes
El-Kébir Boukas

Nonlinear and Adaptive Control with Applications
Alessandro Astolfi, Dimitrios Karagiannis and Romeo Ortega

Stabilization, Optimal and Robust Control
Aziz Belmiloudi

Control of Nonlinear Dynamical Systems
Felix L. Chernous'ko, Igor M. Ananievski and Sergey A. Reshmin

Periodic Systems
Sergio Bittanti and Patrizio Colaneri

Discontinuous Systems
Yury V. Orlov

Constructions of Strict Lyapunov Functions
Michael Malisoff and Frédéric Mazenc

Controlling Chaos
Huaguang Zhang, Derong Liu and Zhiliang Wang

Stabilization of Navier-Stokes Flows
Viorel Barbu

Distributed Control of Multi-agent Networks
Wei Ren and Yongcan Cao

Alexandre Sanfelice Bazanella •
Lucíola Campestrini • Diego Eckhard

Data-Driven Controller Design

The H_2 Approach

Springer

Alexandre Sanfelice Bazanella
Universidade Federal do Rio Grande do Sul (UFRGS)
Porto Alegre
Rio Grande do Sul
Brazil
bazanela@ece.ufrgs.br

Lucíola Campestrini
Universidade Federal do Rio Grande do Sul (UFRGS)
Porto Alegre
Rio Grande do Sul
Brazil

Diego Eckhard
Universidade Federal do Rio Grande do Sul (UFRGS)
Porto Alegre
Rio Grande do Sul
Brazil

ISSN 0178-5354 Communications and Control Engineering
ISBN 978-94-017-8356-9 ISBN 978-94-007-2300-9 (eBook)
DOI 10.1007/978-94-007-2300-9
Springer Dordrecht Heidelberg London New York

Printed on acid-free paper

Springer is part of Springer Science+Business Media (www.springer.com)

To Adriane, the light and the warmth of my life.

Alexandre Sanfelice Bazanella

To Thiago and Lourdes, my greatest supporters.

Lucíola Campestrini

For Priscila, my joy and inspiration.

Diego Eckhard

Preface

Since the early 1990's a variety of control design methods that are constructed directly upon batches of input-output data collected from the process to be controlled have appeared in the literature. These methods contrast with model-based control design mainly in two fundamental aspects: they are not based on the knowledge of a process model and they do not intend to freely determine the controller's transfer function. Instead, they make direct use of the information carried by the measured data in order to adjust the numerical parameters of a controller whose transfer function has a previously specified and fixed structure. Accordingly, these methods became known as *data-based*, in opposition to model-based, or alternatively as *data-driven* methods.

This same challenge of adjusting the parameters of a controller whose transfer function structure is given a priori, without ever obtaining a model for the process, has been undertaken in the context of adaptive control, since at least the 1960's. This has become known as the *direct approach* to adaptive control, in contrast with the *indirect approach*, in which the controller's parameters are adjusted through a model-based design, which is performed after the identification of the parameters of a process model and with the application of the certainty equivalence principle. Adaptive control has been a major field of research in control theory ever since, and colossal amounts of literature and successful applications of adaptive control have been developed over the past half-century. However, most quotidian industry applications do not seem to have assimilated this evolution. This gap between practical industry applications and the adaptive control theory, along with extraneous nonlinear behavior introduced by the adaptation mechanisms, propelled a surge of interest in the data-driven alternative for controller's adaptation, a surge that has been gaining momentum since its onset.

Attempts to delimit exactly the borders between model-based control, adaptive control, and data-driven control would most likely be unproductive; instead, let us propose a rather elastic definition of data-driven control. The term data-driven (or data-based) control refers to the methodologies whose aim is to design the parameters of a fixed-structure controller based on a reasonably large batch of input-output data, without any attempts to perform control design based on a process model.

Data-driven control is thus different from model-based control because the design is not based on a process model, even though approximative process models can be used for secondary purposes in some data-driven control methods. And data-driven control differs from adaptive control essentially by the fact that parameter adjustments are always based on large batches of data rather than on a single input-output sample or a few samples, as is usual in adaptive control. This one difference between data-driven control design and adaptive control has major implications, both theoretical and practical, and yet a substantial part of the theory in this book can be applied to direct adaptive control as well.

Some conceptually distinct data-driven approaches to control design appear in the literature. Yet, it seems fair to say that most of the methodologies are built around one of the most familiar concepts in control systems theory: the optimization of a performance criterion, where performance is measured by the H_2 norm of a particular signal in the loop. Representative of these methodologies are the pioneering works of Hjalmarsson, Gunnarsson and Gevers [3], Kammer, Bitmead and Bartlett [4], Campi, Lecchini and Savaresi [2], Karimi, Mišković and Bonvin [5], and Shi and Skelton [6]. These pioneering works later developed into sound design methodologies, as well as into applications, described in many other papers which constituted fundamental sources for the writing of this book. Each one of these methodologies has been baptized by their authors, respectively as: Iterative Feedback Tuning (IFT), Frequency Domain Tuning (FDT), Virtual Reference Feedback Tuning (VRFT), Correlation-based Tuning (CbT) and Markov LQG Control.

This is what this book is about: it intends to present a comprehensive analysis of this H_2 approach to data-driven control design, providing a common theoretical framework to these methodologies that have been presented separately in the literature since the early 1990's. This common theoretical framework also fits a large family of adaptive control methodologies, like Minimum Variance and Model Reference Adaptive Control. From this unified framework a number of shared properties become apparent, and solutions to shared problems emerge. This unification effort, which has been initiated in [1], is based on the analysis of the problem itself, namely the objective function(s) being minimized, the features of the data used for this purpose and the potential ways of performing the minimization, rather than on the algorithmic details of each particular solution.

The book is primarily intended for PhD students and researchers, whether senior or junior, in control engineering. It should serve as reference material for PhD theses, as well as teaching material for data-driven and adaptive control courses at the graduate level. We hope it will also be useful for advanced engineers willing to apply data-driven control design, by providing them with an understanding of the strengths and limitations of the existing data-driven methodologies and guidelines that will help coding these methods; but it is not a "how-to book".

The book starts with a formal delimitation of the problem and the class of systems treated. General definitions appear in Chap. 1, whereas the formal statement of the H_2 design problem is given in Chap. 2. Also in Chap. 2 is a presentation of the basic properties of the H_2 design problem and a discussion about the different control objectives and what are the designer's choices in choosing her/his performance criterion.

Once the designer has chosen the performance criterion, it must be minimized, which will be done using only input-output data collected from the system. It is possible in many situations to perform this minimization in only "one-shot", that is, with only one batch of data collected in only one operating condition. These "one-shot" solutions, which are the most convenient, are the subject of Chap. 3. In many other situations, however, it is necessary to resort to iterative procedures in which each iteration requires collecting more data, each time with a different controller in the loop. This may be necessary because the theoretical conditions required by the "one-shot" solutions are not met, or because operational constraints of the process require that only small adjustments to existing parameter settings can be made (a very common practical constraint). Accordingly, the subject of iterative optimization occupies the remaining of this book.

In Chap. 4 a general review of optimization is given, in order to set the stage for the chapters to follow. Then, starting in Chap. 5, the particularities of the H_2 cost functions minimized in data-driven control start being explored, bearing in mind that convergence to the globally optimal controller is sought. A number of properties of these particular cost functions and of some basic optimization algorithms when applied to them are presented in Chap. 5, along with guidelines for the optimization. Then, in Chap. 6, the *cost function shaping* concept is presented. Cost function shaping is the name we have given to a set of procedures and maneuvers that change the cost function so that it is more amenable to optimization.

Performing the optimization requires calculation of the cost function's derivatives, and this must be done only with the data collected from the system—no analytical expressions are available. It is only in Chap. 7 that this computing aspect—the calculation of these quantities—is discussed. In this chapter, three different methods are described in some detail and interpreted under the light of the theory presented in the previous chapters: IFT, FDT and CbT.

All along the book, simulation examples are presented that aim to illustrate and explain the concepts presented. These simulation studies do not have the purpose of demonstrating the practical application of these concepts, a task that is accomplished separately, in Chap. 8. There, numerous experimental results showing the data-driven design of controllers for processes of different natures are presented. These designs illustrate how the theory translates into the real world, showing that the methodologies are indeed effective, what are the designer's choices, and how he/she should make these choices taking into account the theoretical concepts presented in this book to obtain the best result from a real data-driven control design.

References

1. A.S. Bazanella, M. Gevers, L. Mišković, B.D.O. Anderson, Iterative minimization of H_2 control performance criteria. Automatica **44**(10), 2549–2559 (2008)
2. M.C. Campi, A. Lecchini, S.M. Savaresi, Virtual reference feedback tuning: A direct method for the design of feedback controllers. Automatica **38**, 1337–1346 (2002)

3. H. Hjalmarsson, S. Gunnarsson, M. Gevers, A convergent iterative restricted complexity control design scheme, in *33rd IEEE Conference on Decision and Control*, Lake Buena Vista, USA, 1994
4. L.C. Kammer, R.R. Bitmead, P.L. Bartlett, Direct iterative tuning via spectral analysis. Automatica **36**, 1301–1307 (2000)
5. A. Karimi, L. Mišković, D. Bonvin, Iterative correlation-based controller tuning. Int. J. Adapt. Control Signal Process. **18**, 645–664 (2004)
6. G. Shi, R.E. Skelton, Markov data-based LQG control. J. Dyn. Syst. Meas. Control **122**(3), 551–560 (2000)

Porto Alegre, Brazil

Alexandre Sanfelice Bazanella
Lucíola Campestrini
Diego Eckhard

Acknowledgements

Most of the funding for our research has been provided, through various grants, by CAPES and CNPq—respectively the Coordination for the Advancement of Higher Level Personnel from the Brazilian Ministry of Education and the National Council for Scientific and Technological Development from the Brazilian Ministry of Science and Technology. This continuous and dependable support has been, and still is, fundamental to our work.

The first and second authors have also received greatly appreciated support from the Belgian Programme on Interuniversity Attraction Poles (PAI), initiated by the Belgian Federal Science Policy Office, during their stays at the Université Catholique de Louvain. During those periods and beyond, Michel Gevers has been a major source of inspiration, as well as an enthusiastic working partner; we are deeply grateful and honored for this collaboration. Also very important has been the collaboration of the first author with Brian D.O. Anderson—another fundamental influence—and Ljubiša Mišković.

Last, but not least, we thank the various authors that have developed the methods and concepts that we have gathered up in this book. These giants on whose shoulders we have stood are acknowledged in the reference list.

Contents

1 **Definitions** 1
1.1 The Process 1
1.2 The Control 3
1.3 The Closed-Loop System 5
1.4 The Design Problem 6
References 6

2 **H_2 Performance Criteria** 7
2.1 Introduction 7
2.2 The Different Criteria 8
2.2.1 Reference Tracking—The Model Reference Control 8
2.2.2 Noise Rejection—The Minimum Variance Control 9
2.2.3 The Composite Performance 10
2.2.4 Economy of Control Effort 11
2.3 Duality with System Identification—The Ideal Controllers 12
2.3.1 Reference Tracking 12
2.3.2 Noise Rejection 16
2.3.3 The Composite Criterion 18
2.4 Beware of What You Ask for—Choosing the Reference Model 19
2.4.1 Too Ambitious Performance 21
2.5 Chapter Conclusions 24
References 25

3 **One-Shot Optimization—The VRFT Method** 27
3.1 Introduction 27
3.2 The Ideal Case 27
3.2.1 Generation of the Virtual Reference—A Caveat 32
3.3 The Mismatched Case 34
3.4 Dealing with Non-minimum Phase Plants 39
3.4.1 The Flexible VRFT Criterion 41
3.4.2 Implementation Issues 43
3.4.3 Two-Step Procedure 44

3.5 The Noisy Case . . . 45
3.5.1 Choosing the Instrumental Variable . . . 47
3.6 Case Studies . . . 49
3.6.1 The Need of Instrumental Variables . . . 50
3.6.2 Applying the Flexible Criterion . . . 54
3.7 Chapter Conclusions . . . 65
References . . . 66

4 Iterative Optimization . . . 69
4.1 Some Things to Remember from Calculus . . . 69
4.2 Optimization Algorithms (Dynamic Systems) . . . 72
4.2.1 Autonomous Systems . . . 75
4.3 The Basic Algorithms and Their Convergence . . . 75
4.3.1 Steepest Descent . . . 75
4.3.2 Other Search Directions . . . 80
4.3.3 Newton-Raphson . . . 81
4.3.4 Robustness . . . 85
4.4 Chapter Conclusions . . . 87
References . . . 88

5 Convergence to the Globally Optimal Controller . . . 89
5.1 J_y—The Reference Tracker . . . 90
5.2 J_e—The Variance Minimizer . . . 102
5.3 The Mismatched Case . . . 104
5.4 Choosing the Algorithm Parameters . . . 108
5.4.1 The Search Direction . . . 108
5.4.2 The Step Sizes—First Solution . . . 109
5.4.3 The Step Sizes—Second Solution . . . 110
5.5 A Case Study . . . 112
5.6 Chapter Conclusions . . . 115
References . . . 117

6 Cost Function Shaping . . . 119
6.1 Introduction . . . 119
6.2 The Problem Data . . . 119
6.3 Cautious Control . . . 121
6.4 Manipulation of the Reference Spectrum . . . 124
6.4.1 Properties of the Sensitivity . . . 125
6.4.2 Applying a Different Reference . . . 127
6.4.3 Windowing . . . 129
6.5 Case Studies . . . 133
6.5.1 A PID Design . . . 133
6.5.2 A Noisy Process . . . 138
6.6 Chapter Conclusions . . . 142
References . . . 144

7 Computations . . . 145
7.1 Iterative Feedback Tuning . . . 146
7.1.1 Derivation . . . 146
7.1.2 Extensions . . . 150
7.2 Frequency Domain Tuning . . . 154
7.2.1 Derivation . . . 155
7.2.2 Extensions . . . 158
7.3 Correlation-Based Tuning . . . 159
7.3.1 Derivation . . . 160
7.4 Chapter Conclusions . . . 163
References . . . 163

8 Experimental Results . . . 165
8.1 A Liquid Flow Process . . . 166
8.1.1 Direct Method . . . 169
8.1.2 Iterative Method . . . 174
8.2 A DC Motor . . . 177
8.2.1 Direct Method . . . 178
8.2.2 Iterative Method . . . 186
8.3 A Temperature Process . . . 187
8.3.1 The Favorable Scenario . . . 187
8.3.2 Tough Case . . . 194
References . . . 207

Abbreviations

Acronyms

1DOF	One Degree of Freedom
2DOF	Two Degrees of Freedom
BIBO	Bounded-Input Bounded-Output
CbT	Correlation-based Tuning
DC	Direct Current
DOA	Domain of Attraction
FDT	Frequency Domain Tuning
IFT	Iterative Feedback Tuning
IV	Instrumental Variable
GN	Gauss-Newton
LI	Linearly Independent
LP	Linearly Parameterized
LQG	Linear Quadratic Gaussian
LQR	Linear Quadratic Regulator
LS	Least Squares
LTI	Linear Time Invariant
MR	Model Reference
MRAC	Model Reference Adaptive Control
MV	Manipulated Variable
NMP	Non-Minimum Phase
NR	Newton Raphson
PE	Persistently Exciting
PI	Proportional-Integral
PID	Proportional-Integral-Derivative
PRBS	Pseudo Random Binary Signal
PV	Process Variable
SD	Steepest Descent
SISO	Single-Input Single-Output
SNR	Signal-to-Noise Ratio
SPR	Strictly Positive Real

SRp	Sufficiently Rich of order p
VRFT	Virtual Reference Feedback Tuning

Notation and Symbols

$E[\cdot]$	Expected value of a random variable or of a stationary stochastic process
$\bar{E}[x(t)]$	$\lim_{N\to\infty} \frac{1}{N} \sum_{t=1}^{N} E[x(t)]$
z	Forward time-shift operator, defined as $zx(t) = x(t+1)$
$\mathbb{R}$	The set of real numbers
$\mathbb{R}^+$	The set of positive real numbers
$\Re\{\cdot\}$	Real part of a complex quantity
$\Im\{\cdot\}$	Imaginary part of a complex quantity
$R_x(\tau)$	$\bar{E}[x(t)x(t-\tau)]$
$\Phi_x(e^{j\omega})$	The spectrum of a stochastic process $x(t)$: $\Phi_x(e^{j\omega}) = \sum_{\tau=-\infty}^{\infty} R_x(\tau)e^{-j\omega\tau}$
I_n	The identity matrix of dimension n
$\min_x f(x)$	Minimization of $f(x)$ where the decision variable is x
$\arg\min f(x)$	Value of x that minimizes $f(x)$
$\triangleq$	The left hand side is defined by the right hand side
$\lvert F(e^{j\omega})\rvert^2$	$F(e^{j\omega})F^*(e^{j\omega})$, where the superscript $*$ means complex conjugate transpose
$\mathrm{var}_{\hat{x}}$	Variance of the estimate of x
$\nabla J(\rho)$	Gradient of the function $J(\rho)$ with respect to ρ
$\nabla^2 J(\rho)$	Hessian of the function $J(\rho)$ with respect to ρ
$\lVert\cdot\rVert$	2-norm $\lVert\cdot\rVert_2$
$x_L(t)$	Filtered version of the signal $x(t)$ at time t
$\{x(t)\}_{t=1,\dots,N}$	Vector containing N samples of the signal $x(t)$
$\hat{\alpha}$	The estimate of α
$\angle x$	The argument (phase) of the complex number x
$\mathrm{dist}(\rho^1, \rho^2)$	$\max_\omega \lvert\angle S(e^{j\omega}, \rho^1) - \angle S(e^{j\omega}, \rho^2)\rvert$
$\mathrm{sign}(\cdot)$	The sign function: $\mathrm{sign}(x) = 1$ if $x > 0$ and $\mathrm{sign}(x) = -1$ if $x < 0$
$sq\left(\frac{2\pi}{T}t\right)$	A square wave with period T
N	Size of the data-set
$r(t)$	Reference signal at time t
$y(t)$	Process' output variable at time t
$u(t)$	Control input variable at time t (controller's output)
$v(t)$	Process noise at time t
$e(t)$	Zero-mean white noise process
σ_e^2	Variance of the white noise process
$y_d(t)$	Desired closed-loop system response at time t
$\bar{r}(t)$	Virtual reference signal at time t
$\bar{e}(t)$	Virtual reference tracking error at time t
$\varphi(t)$	Regressor vector at time t
$\zeta(t)$	Instrumental variable at time t
$G(z)$	Process' transfer function
$H(z)$	Noise's transfer function

$T_d(z)$	Reference model for the reference tracking problem
$S_e(z)$	Desired sensitivity function for the minimum variance control problem
$T_e(z)$	Desired complementary sensitivity function for the noise rejection problem
$C_0(z)$	Initial controller
ρ	Real vector used to parameterize the controller; dimension p
$C(z, \rho)$	Controller's transfer function parameterized in terms of ρ
$T(z, \rho)$	Closed-loop transfer function obtained with $C(z, \rho)$—complementary sensitivity
$S(z, \rho)$	Closed-loop transfer function obtained with $C(z, \rho)$—sensitivity
$y_r(t, \rho)$	Closed-loop system output at time t neglecting the effect of noise
$y_e(t, \rho)$	Closed-loop system output at time t due to the noise alone
$\varepsilon(t, \rho)$	$y_d(t) - y(t, \rho)$ The error between the desired and the achieved closed-loop response at time t
$C_d(z)$	Ideal controller for the reference tracking problem
ρ_d	Ideal parameter vector, for which $C(z, \rho_d) = C_d(z)$
$C_e(z)$	Ideal controller for the noise rejection problem
ρ_*	The global minimum of a cost function
$K(z)$	Difference between the ideal controller and the best controller allowed by the controller class
ρ_0	Parameter vector of the initial controller
ρ_m	Mean value of the parameter ρ obtained from Monte Carlo runs
ρ^{IV}	Estimate of ρ through instrumental variables
ρ_i	Parameter vector ρ obtained at iteration i
ρ^i	Instance i of the parameter vector ρ
ϱ_i	Element i of the parameter vector ρ, when ρ has more than one element
$\bar{C}(z)$	Vector of rational functions which forms a linear parameterized controller with ρ
$\mathcal{C}$	Controller class
$\mathcal{C}_{PID}$	Controller class of PID controllers
$\mathcal{C}_{PI}$	Controller class of PI controllers
k_p, k_i and k_d	Proportional, integral and derivative gains, respectively
$T(z, \eta)$	Flexible reference model
η	Parameter vector of the flexible reference model; dimension q
ϑ_i	Element i of the parameter vector η
(η_*, ρ_*)	Global minimum for the flexible criterion $J_0^{VR}(\rho)$
ρ_i^{IV}	Parameter vector ρ obtained using IV in iteration i
η_i^{IV}	Parameter vector η obtained using IV in iteration i
$\eta_{i,m}^{IV}$	Mean value of the parameter vector η obtained from Monte Carlo runs in iteration i using IV
$C_{VR}(\rho)$	Controller obtained with the application of the VRFT method
$C_{LS}(\rho)$	Controller obtained with the application of the VRFT method, using least squares

$C_{IV}(\rho)$	Controller obtained with the application of the VRFT method, using instrumental variables
$J(\cdot)$	Cost function
$J_y(\rho)$	Reference tracking performance criterion
$J_e(\rho)$	Noise rejection performance criterion
$J_T(\rho)$	$J_y(\rho) + J_e(\rho)$
$J_e^r(\rho)$	Relaxed noise rejection criterion
$J_u(\rho)$	Control effort performance criterion
$J_\lambda(\rho)$	Weighted sum of the performance criterion and the control effort
λ	Design weight parameter
$J_c(\rho)$	Performance criterion of the CbT method
$J^{VR}(\rho)$	Performance criterion of the VRFT method
$J_0^{VR}(\rho)$	Flexible cost criterion for the VRFT method
$J_{IFT}(\rho)$	Performance criterion of the IFT method
$J_{FDT}(\rho)$	Performance criterion of the FDT method
$J_S(\rho)$	$J_T(\rho)\|_{T_d(z)=1}$
$\hat{J}_T(\rho)$	$\frac{1}{N}\sum_{t=1}^{N}(y(t,\rho) - y_d(t))^2$—$J_T(\rho)$ computed from data
$\check{J}_T(\rho)$	$\sqrt{\hat{J}_T(\rho)}$
$\mathcal{B}_\epsilon(\rho^1)$	$\{\rho : \ \|\rho - \rho^1\|^2 < \epsilon\}$ A ball centered in ρ^1 with radius equal to $\sqrt{\epsilon}$
Γ	The set of all the controller's parameter values that render the closed-loop BIBO-stable
μ_i	Effective step size at iteration i
γ_i	Gradient factor to scale the step size at iteration i
$V(\cdot)$	Lyapunov function
$\Sigma(z,\rho)$	Sensitivity matrix
$b_{C(z,\rho)}$	$\|\Sigma(z,\rho)\|_\infty^{-1}$ Stability margin
$\mathbf{0}$	Vector of zeros of appropriate dimension

Chapter 1
Definitions

1.1 The Process

In this book we consider the control of linear time-invariant (LTI) discrete-time single-input-single-output (SISO) processes. The processes under consideration can be described by (1.1) below and the technical assumptions that follow, which are necessary to specify precisely the problem at hand and also to guarantee that it is well-posed

$$y(t) = G(z)u(t) + \nu(t). \tag{1.1}$$

In (1.1) $y(t)$ is the process output, $u(t)$ is the control input and $\nu(t)$ is the process noise; z is the forward time-shift operator, defined as $zx(t) = x(t+1)$, and $G(z)$ is the process' transfer function. The output of the process is also referred to as the controlled variable, or simply the process variable (PV). The control input $u(t)$ can be, in principle, freely manipulated; accordingly it is often referred to as the manipulated variable (MV). The process noise $\nu(t)$ represents all stochastic effects that are not captured by the input-output relationship between $u(t)$ and $y(t)$: stochastic disturbances, measurement errors, etc. This noise is assumed to result from LTI filtering of white noise, that is, $\nu(t) = H(z)e(t)$ where $H(z)$ is a transfer function and $e(t)$ is white noise, whose variance will be denoted σ_e^2.

Both the process transfer function $G(z)$ and the noise transfer function $H(z)$ are rational transfer functions. They are also proper, that is, the systems they represent are both causal. It is further assumed, without loss of generality, that $H(\infty) = 1$, that is the impulse response $h(t)$ of the filter $H(z)$ satisfies $h(0) = 1$. Moreover, the noise filter $H(z)$ is BIBO-stable. This setting is standard in control systems theory in general and in adaptive control in particular, as well as in the parameter identification framework.

The process (1.1) is controlled by a linear time-invariant controller, whose transfer function has a given (user specified) structure, parameterized in terms of a real parameter vector $\rho \in \mathbb{R}^p$. That is, the control action $u(t)$ can be written as

$$u(t) = C(z, \rho)(r(t) - y(t)) \tag{1.2}$$

A. Sanfelice Bazanella et al., *Data-Driven Controller Design*,
Communications and Control Engineering,
DOI 10.1007/978-94-007-2300-9_1, © Springer Science+Business Media B.V. 2012

where $r(t)$ is the reference signal and $C(z, \rho)$ is the controller's transfer function, both of which are further specified in the sequel.

The reference $r(t)$ is usually a deterministic signal, whereas the system noise is stochastic, a dichotomy which can cause some technical difficulties in the analysis. This dichotomy is eliminated by the use of a common framework for stochastic processes and deterministic signals in which all are assumed to be quasi-stationary processes—see the book by Ljung [3] for details.

A stochastic process—say $x(t)$—is said to be quasi-stationary if its mean and autocorrelation are both bounded and if the following limit exists:

$$\lim_{N\to\infty} \frac{1}{N} \sum_{t=1}^{N} E[x(t)x(t-\tau)] \triangleq \bar{E}[x(t)x(t-\tau)] \quad \forall \tau$$

where $E[\cdot]$ denotes expectancy and we have defined the notation $\bar{E}[\cdot]$.

For two quasi-stationary processes $x(t)$ and $y(t)$ we can define also the following two quantities

$$R_x(\tau) \triangleq \bar{E}[x(t)x(t-\tau)]$$

and

$$R_{xy}(\tau) \triangleq \bar{E}[x(t)y(t-\tau)]$$

which will be referred to, with some abuse of nomenclature, the auto-correlation of $x(t)$ and the cross-correlation between $x(t)$ and $y(t)$, respectively. The spectrum of a quasi-stationary process and the cross-spectrum between two processes are defined as the discrete Fourier transforms of the corresponding correlations, that is:

$$\Phi_x(e^{j\omega}) \triangleq \sum_{\tau=-\infty}^{\infty} R_x(\tau)e^{-j\omega\tau}$$

and

$$\Phi_{xy}(e^{j\omega}) \triangleq \sum_{\tau=-\infty}^{\infty} R_{xy}(\tau)e^{-j\omega\tau}.$$

A direct consequence of the above definitions is the following property

$$\bar{E}[x^2(t)] = R_x(0) = \frac{1}{2\pi} \int_{-\pi}^{\pi} \Phi_x(e^{j\omega})d\omega \tag{1.3}$$

which is often referred to as Parseval's Theorem; this property will be used many times in this book.

All along this book, both the reference and the noise are assumed to be quasi-stationary. Moreover, they are also assumed to be uncorrelated, that is

$$R_{re}(\tau) = \bar{E}[r(t)e(t-\tau)] = 0 \quad \forall \tau.$$

1.2 The Control

The set of all the controllers that can be implemented with the specified controller structure is called the controller class $\mathcal{C}$:

$$\mathcal{C} = \{C(z, \rho) : \rho \in \mathcal{P}\}$$

where $\mathcal{P} \subseteq \mathbb{R}^p$ is a set of admissible values for the controller parameter vector ρ. Whenever $\mathcal{P}$ is not specified in a given problem, it is assumed that $\mathcal{P} = \mathbb{R}^p$. The controller class $\mathcal{C}$ must be such that the loop transfer function $C(z, \rho)G(z)$ has positive relative degree for all $C(z, \rho) \in \mathcal{C}$, so that the feedback connection is always well-posed; equivalently, the closed loop is not delay-free.

As an example of a controller structure, take the following PID (proportional-integral-derivative) controller.

$$C(z, k_p, k_i, k_d) = k_p + k_i \frac{z}{z-1} + k_d \frac{z-1}{z} \tag{1.4}$$

where k_p, k_i and k_d are the proportional, integral and derivative gains respectively. Setting these gains in a PID control structure like this is by far the most common control engineering task found in practice, as the vast majority of control loops in industry are fixed structure PIDs. The expression (1.4) can also be written more compactly in vector form

$$C(z, \rho) = \rho^T \bar{C}(z) \tag{1.5}$$

where the parameter vector ρ and the parameter independent vector of transfer functions $\bar{C}(z)$ are defined as

$$\rho = \begin{bmatrix} k_p \\ k_i \\ k_d \end{bmatrix} \qquad \bar{C}(z) = \begin{bmatrix} 1 \\ \frac{z}{z-1} \\ \frac{z-1}{z} \end{bmatrix}.$$

This parameterization is linear in the parameter ρ. Most industrial controllers are linearly parameterized in this way, like the PID example above, and this linearity also makes the resulting design problem more amenable to analysis. In addition, any ρ-dependent transfer function can be approximated by a transfer function of this form. Indeed, one can always choose a set of basis functions such that a transfer function $F(z)$ can be represented as

$$F(z) = \sum_{i=1}^{\infty} \alpha_i f_i(z) \tag{1.6}$$

where $\alpha_i \in \mathbb{R}$ and $f_i(z)$ are the basis transfer functions. One example of a celebrated choice for this basis are the Laguerre functions:

$$f_i(z) = \frac{\sqrt{1-a^2}}{(z-a)} \left[\frac{1-az}{z-a}\right]^{i-1}$$

where $a \in (-1, 1)$. A truncation of the series in (1.6) yields

$$F(z) \cong \sum_{i=1}^{p} \alpha_i f_i(z)$$

which can approximate the transfer function to any degree of accuracy desired by choosing sufficiently large p.

For these practical and theoretical reasons, assuming that the controller's transfer function is linearly parameterized does not represent a relevant loss of generality. Accordingly, most (but not all) results in this book are derived for linear parameterizations of this form, so it is worth stating formally this assumption that will be so frequently used.

Assumption LP The transfer function of the controller can be written as

$$C(z, \rho) = \rho^T \bar{C}(z) \tag{1.7}$$

with $\bar{C}(z)$ a vector of p rational transfer functions independent of ρ.

It is important not to confuse this linear parameterization of the controller's transfer function from another form of linear parameterization commonly found in the literature, which is the linear parameterization of the control action:

$$u(t) = \rho^T \phi(t) \tag{1.8}$$

where $\phi(t)$ is a "regressor vector", containing past values of the control signal $u(t)$ and of the tracking error signal $r(t) - y(t)$.

Once the controller structure is fixed, which under Assumption LP amounts to fix the vector $\bar{C}(z)$, the controller design boils down to the setting of the parameter vector ρ; this is often called controller "tuning".

Note that the same controller class admits different parameterizations. For instance, we could parametrize a PI controller $C(z, \rho) = k_p + k_i \frac{z}{z-1}$ as

$$\rho \triangleq \begin{bmatrix} k_p \\ k_i \end{bmatrix} \qquad \bar{C}(z) \triangleq \begin{bmatrix} 1 \\ \frac{z}{z-1} \end{bmatrix}$$

or as

$$\rho \triangleq \begin{bmatrix} \varrho_1 \\ \varrho_2 \end{bmatrix} \qquad \bar{C}(z) \triangleq \begin{bmatrix} \frac{1}{z-1} \\ \frac{z}{z-1} \end{bmatrix}.$$

The class of controllers represented by each one of these two parameterizations is the same, that is, any controller that can be represented in one form can also be represented in the other form. One could also chose the parameterization for this same controller class as

$$\rho \triangleq \begin{bmatrix} \varrho_1 \\ \varrho_2 \\ \varrho_3 \end{bmatrix} \qquad \bar{C}(z) \triangleq \begin{bmatrix} \frac{z+1}{z-1} \\ \frac{z}{z-1} \\ 1 \end{bmatrix}$$

which has three parameters instead of only two; this is not a smart choice, however. It is important, and a necessary condition for the design problem to have a unique solution, that the controller class is represented by the minimal number of parameters possible, which will be called a minimal parameterization. Representing the controller class with a minimal parameterization is something that is desirable to do in the first place, and additional motivations to do it will appear along this book.

A linear parameterization will be minimal if the rational functions forming the basis vector $\bar{C}(z)$ are linearly independent over the field of real numbers, that is, if there is no nonzero vector $\alpha \in \mathbb{R}^p$ such that

$$\alpha^T \bar{C}(z) \equiv 0. \tag{1.9}$$

Notice that the third parameterization proposed above is not minimal; the vector $\alpha = [1\ -2\ \ 1]^T$ satisfies (1.9). Suppose that the rational functions forming the basis vector $\bar{C}(z)$ are Linearly Dependent over the field of real numbers, that is, there exists a nonzero vector $\alpha \in \mathbb{R}^p$ satisfying (1.9). Then, for any scalar c,

$$C(z, \rho + c.\alpha) = (\rho + c.\alpha)\bar{C}(z) = C(z, \rho)$$

that is, for any ρ there exist an affine space for which all the parameter values result in the same controller's transfer function. In order to avoid this, most of the developments in this book will assume that the parameterization is minimal, as formally stated in the following assumption.

Assumption LI The parameterization of the controller is minimal, that is, the elements of $\bar{C}(z)$ are Linearly Independent functions over the field of real numbers:

$$\nexists \alpha \neq \mathbf{0}: \quad \alpha^T \bar{C}(z) \equiv 0. \tag{1.10}$$

1.3 The Closed-Loop System

The system (1.1)–(1.2) in closed loop is represented in Fig. 1.1. The resulting closed-loop relationship between the input signals $r(t)$ and $e(t)$ and the output is as follows

$$\begin{aligned} y(t, \rho) &= T(z, \rho)r(t) + S(z, \rho)H(z)e(t) \qquad (1.11) \\ S(z, \rho) &= \frac{1}{1 + C(z, \rho)G(z)} \\ T(z, \rho) &= \frac{C(z, \rho)G(z)}{1 + C(z, \rho)G(z)} = C(z, \rho)G(z)S(z, \rho). \end{aligned}$$

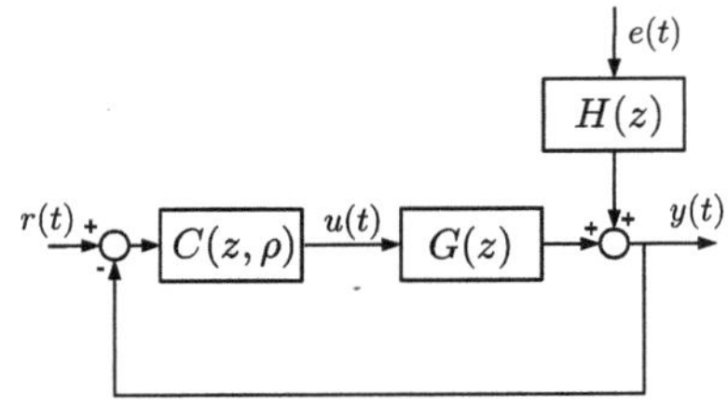

Fig. 1.1 The basic feedback control configuration

In (1.11) we have now made the dependence on the controller parameter ρ explicit in the output signal $y(t, \rho)$. The closed-loop transfer functions $S(z, \rho)$ and $T(z, \rho)$ are called respectively the sensitivity and the complementary sensitivity.

Stability being the primary concern in the control design, it is also useful to define a notation for the set of all controller parameter values that render the closed-loop system BIBO-stable; this set is designated Γ:

$$\Gamma \triangleq \{\rho \in \mathbb{R}^p : T(z, \rho) \text{ is BIBO-stable}\}.$$

1.4 The Design Problem

In summary, the problem studied in this book can be posed as follows.

The Data-Driven Control Design Problem Given a controller structure, which under Assumption LP is specified by a known vector of transfer functions $\bar{C}(z)$, a process with fixed but unknown transfer functions $G(z)$ and $H(z)$ and a performance criterion (to be discussed in the next chapter). Use input-output data $\{u(t), y(t), t = 1, \dots, N\}$ measured from the system to automatically determine the parameter vector ρ that yields the best performance according to this criterion.

The term "automatically" implies that there is no expert human intervention during the process operation. This problem probably qualifies as one of adaptive control, though we prefer to leave this semantic issue aside and refer the reader to the famous paper [1] and the classical book [2], where she/he can find a classical discussion of what is and what is not adaptive control.

References

1. K.J. Åström, Theory and applications of adaptive control – A survey. Automatica **19**(5), 471–486 (1983)
2. K.J. Åström, B. Wittenmark, *Adaptive Control*, 2nd edn. (Addison-Wesley, Reading, 1995)
3. L. Ljung, *System Identification—Theory for the User*, 2nd edn. (Prentice Hall, New York, 1999)

Chapter 2
H_2 Performance Criteria

2.1 Introduction

Control performance criteria are a key element in control systems theory. Not only are they fundamental from a conceptual point of view, but this concept also leads to a huge variety of analysis and design methods, which are formulated as optimization problems. This formulation is, in its general form, expressed as the solution of a problem like

$$\min_{\rho} J(\rho) \tag{2.1}$$

where the controller parameter vector ρ is the decision variable. In (2.1) $J(\rho)$ is an objective function which expresses the system's performance in achieving a prescribed control objective, smaller values of the objective function expressing better performance. The most fundamental control objectives, as well as more sophisticated ones, are quite naturally and effectively expressed as the norm of some signal in the control loop. When the 2-norm

$$\|x(t)\|^2 = \frac{1}{N}\sum_{t=1}^{N}[x(t)]^2$$

is used, this is said to be an H_2 performance criterion.

A large variety of control analysis and design tools have been derived from this general formulation with H_2 performance criteria, since the early days of modern control theory—think of Linear Quadratic Control (LQR/LQG) and Generalized Predictive Control (GPC), for example. These methods can deal with quite sophisticated and ambitious control objectives, but they usually rely on full knowledge of the process and noise, and require the controller's transfer function to be freely chosen. Obtaining a good model for a real process usually demands, among other tasks, collecting data from real system operation. *Data-driven design*, on the other hand, addresses the minimization of the performance criterion (2.1) directly from the data collected from the system, without the intermediate step of deriving a process model

A. Sanfelice Bazanella et al., *Data-Driven Controller Design*,
Communications and Control Engineering,
DOI 10.1007/978-94-007-2300-9_2,

from these data, and for a controller whose structure[1] has been previously fixed—only the parameters in this fixed structure are free to be designed.

In this chapter we will present the H_2 control performance criteria corresponding to the fundamental control objectives: reference tracking, noise rejection and economy of control effort. Most commonly found adaptive control algorithms are concerned with optimizing either reference tracking—like the celebrated Model Reference Adaptive Control—or noise rejection—the equally celebrated Minimum Variance Control. We will dissect each one of these performance criteria, as well as some of its variations and combinations, developing the theoretical framework that will be used in subsequent chapters to analyze the challenges and solutions encountered in their optimization.

2.2 The Different Criteria

2.2.1 Reference Tracking—The Model Reference Control

Reference tracking is concerned with the response of the closed-loop system to the reference alone, disregarding the effect of noise in the output. Let us define this response as

$$y_r(t, \rho) \triangleq T(z, \rho)r(t).$$

A fundamental objective of a control system being to make the process output as close as possible to the reference, the performance from this point of view can be evaluated by the two-norm of the tracking error, that is, by the following performance criterion:

$$J(\rho) = \bar{E}[r(t) - y_r(t, \rho)]^2. \tag{2.2}$$

It is easy to see, and a known fact taught even at the basic levels of control courses, that it is usually impossible to obtain perfect tracking; that is, no controller, whether in the considered class of controllers or not, can make the output to track exactly the reference at all times; the performance criterion in (2.2) can never be made zero. Since perfect tracking is not possible, the tracking objective is usually relaxed into a specification of how close a tracking would be satisfactory for the designer. This is often expressed in terms of control performance measures such as settling time, maximum overshoot, rising time, cutoff frequency, etc. The Model Reference paradigm comes into the scene as an alternative, more detailed, more succinct and analytically treatable description of this relaxation.

In the Model Reference design paradigm, the designer is asked to create a transfer function whose behavior is the one expected from the closed-loop system. This target transfer function is called the *reference model*, and will be henceforth denominated $T_d(z)$—the subscript d standing for "desired". The response desired for the

[1] The $\bar{C}(z)$ vector introduced in the previous chapter.

closed-loop system under a given reference signal $r(t)$ is then $y_d(t) = T_d(z)r(t)$. The response actually obtained in closed-loop from the reference signal[2] is $y_r(t,\rho)$ defined above, which should be as close as possible to $y_d(t)$. Then the controller design consists in finding the controller parameters that make these two signals as close as possible to each other. That is, instead of using the performance criterion in (2.2), the following function $J_y(\rho)$ is defined as the reference tracking performance criterion

$$J_y(\rho) \triangleq \bar{E}[y_r(t,\rho) - y_d(t)]^2 = \bar{E}[(T(z,\rho) - T_d(z))r(t)]^2. \tag{2.3}$$

This control design formulation has been also called, in a more general framework, the model matching control. It can be cast as a linear quadratic regulator (LQR) and thus solved by means of tools such as Riccati equations, Linear Matrix Inequalities (LMI's), Bilinear Matrix Inequalities (BMI's), etc. [3]. Provided, of course, that the process model $G(z)$ is known, which is not the case in this book.

The model reference control design paradigm has been around since at least the 1960's. This paradigm has caught more attention within the adaptive control framework, probably because it lends itself naturally to the automatic adjustment of the controller parameters. Before we move on, let us just note a relevant relationship with another control design paradigm of great success in the adaptive control context: the pole assignment design. In pole assignment, as the name says, the designer assigns the poles of the closed-loop system, that is, the denominator of its transfer function; here we assign the whole transfer function—denominator and numerator.

2.2.2 *Noise Rejection—The Minimum Variance Control*

Another fundamental control objective is to minimize the effect of the noise in the output. The output of the closed-loop system due to the noise alone, that is, disregarding the effect of the reference, is given by

$$y_e(t,\rho) \triangleq S(z,\rho)v(t).$$

The effect of the noise in the output can be measured by the size of this signal, giving rise to the noise rejection performance criterion, which we immediately baptize as $J_e(\rho)$:

$$J_e(\rho) \triangleq \bar{E}[y_e(t)]^2 = \bar{E}[S(z,\rho)v(t)]^2. \tag{2.4}$$

Again, it is clear that no controller in the universe can make the effect of the noise to disappear completely, that is, to make $J_e(\rho) = 0$. Indeed, this would require that $S(z,\rho) = 0\ \forall z$ which in turn would demand $C(z,\rho)G(z) \to \infty\ \forall z$. Since "perfect performance" is not possible for this performance criterion either, a relaxation can

[2] That is, neglecting the effect of noise.

also be thought of here, just as in the case of the reference tracking criterion $J_y(\rho)$ and for the same reasons. To do that, define a *desired* sensitivity function $S_e(z)$, and adopt the following criterion:

$$J_e^r(\rho) \triangleq \bar{E}[(S(z,\rho) - S_e(z))\nu(t)]^2.$$

Even though this relaxation is in all respects dual to the definition of the reference model $T_d(z)$ for the reference tracking criterion, and even though it can lead to better results than using the original $J_e(\rho)$ as the performance criterion, this relaxation is by no means usual and virtually absent from the literature. This fact is probably due to two conceptual differences between the two criteria. First, in tracking the reference, usually the main concern when adjusting the controller's parameters is with the transient performance, as steady-state performance is usually guaranteed by the application of the internal model principle in the choice of the controller's structure.[3] And it is rather easy to come up with a transfer function that provides the desired transient performance. For noise rejection, on the other hand, devising a reasonable sensitivity function may not be as intuitive, particularly taking into account that the noise model $H(z)$ is unknown. Second, the input signals in each case—$J_y(\rho)$ and $J_e(\rho)$—are of different nature. In the noise rejection case, the input signal is the filtered white noise $\nu(t)$, whereas the reference can in principle be anything and is most likely to be deterministic. The random signal $\nu(t)$ being not measurable, as opposed to $r(t)$ involved in the tracking performance criterion, complicates the practical computation of the quantities that would be necessary for its optimization—typically the derivatives of the performance criterion. This issue of computation will become clearer in Chap. 7, where we will see computation schemes that yield the derivatives of $J_e(\rho)$ and $J_y(\rho)$, but not of $J_e^r(\rho)$. For these reasons we will adhere to tradition and consider the performance criterion $J_e(\rho)$ instead of the more generic $J_e^r(\rho)$.

2.2.3 The Composite Performance

Each one of the two performance criteria just presented (reference tracking and noise rejection) represents a conceptually different control objective and each one of them has a theoretical and practical relevance of its own. Numerous control design methods, of data-driven, adaptive and model-based natures, have been developed for each one of them independently. Most commonly found adaptive control algorithms are concerned either with minimizing the reference tracking criterion $J_y(\rho)$ (the celebrated Model Reference Adaptive Control) or with the noise rejection criterion $J_e(\rho)$ (Minimum Variance Control).

But equally important is the following *composite performance criterion* $J_T(\rho)$:

$$J_T(\rho) \triangleq \bar{E}[y(t,\rho) - y_d(t)]^2. \tag{2.5}$$

---

[3]Including integral action in the controller being by far the most common instance.

Notice the difference between the definitions of $J_T(\rho)$ and $J_y(\rho)$: the definition of $J_y(\rho)$ involves the *tracking error* $y_r(t,\rho) - y_d(t)$, disregarding the effect of noise in the output, whereas in $J_T(\rho)$ the *output error* $y(t,\rho) - y_d(t)$ appears. We call it the composite criterion because it takes into account both external signals acting on the system $r(t)$ and $v(t)$, and as a result it equals the sum of the two previous ones, that is:

$$J_T(\rho) = \bar{E}[(T(z,\rho) - T_d(z))r(t)]^2 + \bar{E}[S(z,\rho)v(t)]^2 = J_y(\rho) + J_e(\rho). \quad (2.6)$$

To prove this equality it suffices to develop the composite objective function from its definition (2.5):

$$\begin{aligned} J_T(\rho) &= \bar{E}[T(z,\rho)r(t) + S(z,\rho)v(t) - T_d(z)r(t)]^2 \\ &= \bar{E}[(T(z,\rho) - T_d(z))r(t) + S(z,\rho)v(t)]^2 \\ &= \bar{E}[(T(z,\rho) - T_d(z))r(t)]^2 + \bar{E}[S(z,\rho)v(t)]^2 \\ &\quad + 2\bar{E}[(T(z,\rho) - T_d(z))r(t)S(z,\rho)v(t)] \end{aligned}$$

and to realize that the last term in this sum is zero because the reference and the noise are independent.

2.2.4 Economy of Control Effort

Many H_2 design methodologies, including some data-driven methodologies, consider an additional control objective in their performance criteria, which is the economy of control energy—or effort

$$J_u(\rho) = \bar{E}[u(t)]^2. \quad (2.7)$$

This performance criterion does not make sense if used alone. Minimizing $J_u(\rho)$ above would always result in $u(t) \equiv 0$ (an open-loop system), and we don't need an optimization procedure to tell us that the best choice regarding economy of control energy is to turn the control off. Using this performance criterion in a control design only makes sense when combined with another performance criterion, such as reference tracking or noise rejection or their combination. The performance criterion to be minimized in such cases will be a weighted sum of the just defined $J_u(\rho)$ and whatever performance criterion $J(\rho)$ is to be minimized ($J_T(\rho)$, $J_e(\rho)$ or $J_y(\rho)$):

$$J_\lambda(\rho) = J(\rho) + \lambda J_u(\rho) \quad (2.8)$$

where $\lambda \in \mathbb{R}$ is a design parameter which weighs the relative importance of control economy versus the performance as specified by $J(\rho)$.

Energy saving is among the strongest social and industrial concerns nowadays, and probably even more in the future, but still we decline to include it explicitly in our analysis almost everywhere in the remaining of this book. This is not because we are not concerned about saving control effort whenever possible—quite the contrary—but instead because the addition of this term is not necessary to accomplish this end.

In a control design defined by the performance criterion $J_\lambda(\rho)$ in (2.8) there is an additional parameter to be chosen by the designer: the weight λ given to the control effort. It is not clear how to choose λ in order to obtain the desired effect, and most such formulations rely on trial and error to do so. An equivalent effect on control effort's savings can be obtained by properly choosing the reference model $T_d(z)$. Existing constraints in the control action can be taken into account in the choice of the reference model $T_d(z)$, so that we do not require the controller to do any more effort than necessary to provide an appropriate performance. For instance, it may be desirable and possible to achieve a deadbeat response $T_d(z) = \frac{1}{z^n}$ but, if such a performance would require too much control energy, the designer can settle for a slower time response which is also acceptable, and specify a slower reference model—say $T_d(z) = \frac{1}{(z-0.4)^n}$. This slower reference model will most likely require significantly less control effort. So, relaxing the requirements in $T_d(z)$ has a similar effect to adding the term $J_u(\rho)$ in the performance criterion. There seem to be no quantitative guidelines for choosing either one (the reference model and the weight λ) to this end, so it may be wiser to have just one thing to choose—the reference model—instead of two. Accordingly, in almost everything that follows in this book we will not treat the performance criterion $J_u(\rho)$, which also corresponds to set $\lambda = 0$ in (2.8) in all instances.

2.3 Duality with System Identification—The Ideal Controllers

Our aim in this section is to provide a central concept in the framework for the analysis of the properties of data-driven design that will be presented along this book: the "ideal controller". There are two fundamental objectives to be pursued by the control design: noise rejection and reference tracking. Each control objective is best accomplished individually by a given controller, which is the one that, among all linear time invariant controllers in the universe (and not only within the class, and not even with causality or stability constraints), provides the minimum value for the performance criterion. This is what we call the ideal controller, and there is one ideal controller for each process and for each control objective. We will see in the following that the ideal controller plays in the data-driven control design a very similar role to the one played by the "real system" in system identification, and that this analogy can be used to our benefit.

2.3.1 Reference Tracking

Let us start by analyzing the reference tracking objective function, that is, the performance criterion $J_y(\rho)$ defined previously and reproduced below

$$J_y(\rho) = \bar{E}[(T(z,\rho) - T_d(z))r(t)]^2.$$

To that end, define the ideal controller $C_d(z)$, which is the controller transfer function that would exactly achieve the desired closed-loop transfer function $T_d(z)$:

$$C_d(z) = \frac{T_d(z)}{G(z)(1 - T_d(z))}. \tag{2.9}$$

Should the ideal controller $C_d(z)$ be put in the control loop, the objective function would evaluate to zero—that is, $J_y(\rho) = 0$. The ideal controller may or may not belong to the class of controllers considered. Only when it does the closed-loop system can be made to behave exactly as specified by the reference model by a proper choice of the parameter ρ. Whether this is the case or not is a critical issue in determining the properties of a model reference design. So, let us formalize this assumption.

Assumption $\mathbf{B}_y$ (Matched control) $C_d(z) \in \mathcal{C}$ or, equivalently,

$$\exists \rho_d : C(z, \rho_d) = C_d(z). \tag{2.10}$$

This assumption is quite similar to the standard assumption in identification theory that the process model belongs to the model class considered. Though assumptions of this nature are standard in our context [1, 2], they are not weak ones. We can, however, expect them to be violated only moderately in a well formulated design problem. Indeed, it does not make good sense to formulate a problem in which one searches for a performance that is radically different from what can be achieved. Now, it is important to know, for any particular case, whether or not Assumption B_y is satisfied, which usually requires knowledge of a model class for the process. On the other hand, it is possible to arrange things so that this assumption is satisfied. The following example briefly illustrates these ideas, which will be extensively used in this book.

Example 2.1 Consider that a process is controlled by a PID controller whose transfer function is

$$C(z, \rho) = \frac{\varrho_1(z - \varrho_2)(z - \varrho_3)}{z(z - 1)} \tag{2.11}$$

with $\rho = [\varrho_1\ \varrho_2\ \varrho_3]^T$ as the parameter vector to be adjusted. The performance criterion to be minimized is $J_y(\rho)$ with the following reference model:

$$T_d(z) = \frac{1 - c}{z - c}$$

where $c \in (0, 1)$ is a given constant.

The controller class is defined as the set of all transfer functions with this particular structure defined in (2.11):

$$\mathcal{C}_{PID} = \left\{ C(z, \rho) = \frac{\varrho_1(z - \varrho_2)(z - \varrho_3)}{z(z - 1)} : \rho \in \mathbb{R}^3 \right\}.$$

The job of setting the PID parameters consists in choosing, among all the controllers in the class $\mathcal{C}_{PID}$, the best one according to the performance criterion $J_y(\rho)$. If the ideal controller is among them—that is, if $C_d(z) \in \mathcal{C}_{PID}$—then it is certainly the best choice. Now assume that the process has a second-order BIBO-stable transfer function:

$$G(z) = \frac{kz}{(z-a)(z-b)}$$

with fixed but unknown $a,\ b \in (-1, 1)$, $k \in \mathbb{R}$. Then, using (2.9), the ideal controller is given by

$$\begin{aligned} C_d(z) &= \frac{\frac{1-c}{z-c}}{\frac{kz}{(z-a)(z-b)}(1 - \frac{1-c}{z-c})} \\ &= \frac{\frac{1-c}{z-c}}{\frac{kz}{(z-a)(z-b)}(\frac{z-1}{z-c})} \\ &= \frac{(1-c)}{k}\frac{(z-a)(z-b)}{z(z-1)} \end{aligned}$$

which belongs to the class $\mathcal{C}_{PID}$, with

$$\rho_d = \begin{bmatrix} \frac{1-c}{k} \\ a \\ b \end{bmatrix}.$$

Thus, using only the knowledge of a system class to which the process belongs, it is possible to establish that Assumption B_y is satisfied.

Now assume that the same process is controlled by a PI controller within the following class:

$$\mathcal{C}_{PI} = \left\{ C(z, \rho) = \frac{\varrho_1(z - \varrho_2)}{z-1} : \rho \in \mathbb{R}^2 \right\}$$

where the parameter vector has been defined as $\rho = [\varrho_1\ \varrho_2]^T$. It is evident that $C_d(z) \notin \mathcal{C}_{PI}$; equivalently Assumption B_y is not satisfied. However, if we really want to work under this assumption, we may choose another reference model which provides a response $y_d(t)$ to be tracked that is similar to the original one. It is straightforward to verify that with the second order reference model

$$T_d(z) = \frac{Kz}{z^2 + (K-1-b)z + b}$$

where K is any real number chosen by the designer, $C_d(z) \in \mathcal{C}_{PI}$ and $\rho_d = [\frac{K}{k}\ a]^T$. And if the parameter K is chosen such that the dominant pole of this new reference model is close to the unique pole c of the original reference model, then both yield similar step responses. Figure 2.1 illustrates for an example: $b = 0.24$, $c = 0.8$,

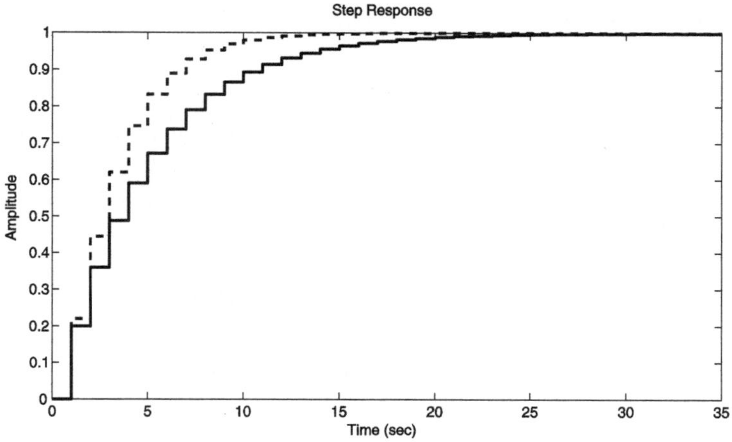

Fig. 2.1 Step responses of the two reference models in Example 2.1: $T_d(z) = \frac{0.2}{z-0.8}$ (continuous line) and $T_d(z) = \frac{0.22z}{z^2-1.02z+0.24}$

$K = 0.22$, and $r(t)$ is a step signal. This new choice of reference model, however, requires the knowledge of the process' pole b.

Assumption B_y is a quite realistic assumption in many cases. Accordingly, we analyze the properties of the solution of the model reference design and derive a number of properties for it under this assumption, which we call the matched control case. We also show in Sect. 2.4 that actually we can—and should—turn the problem around, as in the last part of the example. That is, instead of just picking a reference model and hoping that Assumption B_y is satisfied, we can try and choose the reference model such that it is.

In the mismatched control case, when Assumption B_y is not satisfied, we define the mismatch in terms of the ideal controller. Let $\rho_* = \arg\min J_y(\rho)$; then $C(z, \rho_*)$ is the best controller allowed by the controller class. The mismatch is defined as the difference between this controller and the ideal controller:

$$K(z) \triangleq C_d(z) - C(z, \rho_*). \tag{2.12}$$

This is similar to the bias definition in system identification [4]—another similarity with system identification theory found in this book, and certainly not the last. The norm of the transfer function $K(z)$ can be used as a measure of mismatch. The case where Assumption B_y is not satisfied will also be analyzed, and this analysis will be built upon the results obtained for the matched control case.

When Assumption B_y is satisfied, the global minimum of the performance criterion is ρ_d. On the other hand, when Assumption B_y is not satisfied, the global minimum of $J_y(\rho)$ will be dependent of the reference spectrum. Whether the global minimum is unique or not depends on the richness of the input $r(t)$, in both cases. These properties will be analyzed further ahead in this book.

2.3.2 Noise Rejection

To analyze the noise rejection performance criterion in a similar way to what has just been done for the reference tracking criterion, defining an ideal controller, some preparation is needed. This is due to the fact that, unlike the reference tracking cost, no relaxation has been included in $J_e(\rho)$ and as a result this cost function can not be made zero by any controller.

Applying Parseval's Theorem to (2.4) and using the fact that $v(t) = H(z)e(t)$, and that $e(t)$ is white noise (thus with a flat spectrum) leads to

$$\begin{aligned} J_e(\rho) &= \frac{1}{2\pi}\int_{-\pi}^{\pi} |S(e^{J\omega},\rho)|^2 \Phi_v(e^{J\omega})d\omega \\ &= \frac{1}{2\pi}\sigma_e^2 \int_{-\pi}^{\pi} |S(e^{J\omega},\rho)|^2 |H(e^{J\omega})|^2 d\omega. \end{aligned} \tag{2.13}$$

So $J_e(\rho) = 0$ would require $S(e^{J\omega},\rho) = 0\ \forall\omega$, but no controller can provide this because $S(z,\rho) = \frac{1}{1+C(z,\rho)G(z)}$—an argument already presented a few pages earlier. Hence, in order to check which controller provides the smallest possible value for $J_e(\rho)$, we start by checking which value is this. Manipulating (2.13) yields

$$\begin{aligned} J_e(\rho) &= \sigma_e^2 \frac{1}{2\pi}\int_{-\pi}^{\pi} |1 + H(e^{J\omega})S(e^{J\omega},\rho) - 1|^2 d\omega \\ &= \sigma_e^2 \frac{1}{2\pi}\int_{-\pi}^{\pi} (1 + H(e^{J\omega})S(e^{J\omega},\rho) - 1)(1 + H(e^{J\omega})S(e^{J\omega},\rho) - 1)^* d\omega \\ &= \sigma_e^2 \frac{1}{2\pi}\int_{-\pi}^{\pi} (1 + [H(e^{J\omega})S(e^{J\omega},\rho) - 1] + [H(e^{J\omega})S(e^{J\omega},\rho) - 1]^* \\ &\quad + |H(e^{J\omega})S(e^{J\omega},\rho) - 1|^2) d\omega \\ &= \sigma_e^2 \frac{1}{2\pi}\int_{-\pi}^{\pi} (1 + 2\Re\{H(e^{J\omega})S(e^{J\omega},\rho) - 1\} + |H(e^{J\omega})S(e^{J\omega},\rho) - 1|^2) d\omega. \end{aligned} \tag{2.14}$$

Now observe that, by hypothesis,

$$\lim_{z\to\infty} H(z) = 1$$

and

$$\lim_{z\to\infty} C(z,\rho)G(z) = 0 \quad \forall\rho.$$

So, by construction,

$$\lim_{z\to\infty} S(z,\rho) = \lim_{z\to\infty} \frac{1}{1 + C(z,\rho)G(z)} = 1.$$

Hence $[H(z)S(z,\rho)-1]$ is a strictly proper transfer function, and the integral of the real part of a strictly proper transfer function is zero. As a result, we can write

$$J_e(\rho)=\sigma_e^2+\sigma_e^2\frac{1}{2\pi}\int_{-\pi}^{\pi}|H(e^{j\omega})|^2\left|S(e^{j\omega},\rho)-\frac{1}{H(e^{j\omega})}\right|^2 d\omega. \tag{2.15}$$

Equation (2.15) consists of a constant term and a controller-dependent term which is positive semi-definite. It is seen that the least possible value for the noise rejection term equals σ_e^2. Moreover, it is also seen that this least value is achieved, among all linear time-invariant controllers, regardless of any additional constraints, when the controller is such that the sensitivity equals the inverse of the noise filter, that is, for $S(z,\rho)=\frac{1}{H(z)}$, because then the second term in (2.15) vanishes. This sensitivity is the desired sensitivity regarding this performance criterion, and the corresponding complementary sensitivity plays here a similar role to $T_d(z)$ in the tracking performance criterion, in the sense that it represents the desired closed-loop transfer function from $r(t)$ to $y(t)$. Let us define this desired transfer function:

$$T_e(z)=1-\frac{1}{H(z)}.$$

The controller which achieves this desired closed-loop behavior is given by

$$C_e(z)=\frac{T_e(z)}{G(z)(1-T_e(z))}=\frac{H(z)-1}{G(z)}. \tag{2.16}$$

This is the "ideal controller" regarding the noise rejection performance $J_e(\rho)$, which is usually called "minimum variance" controller in the literature. This controller may very well result in an internally unstable closed-loop, which happens if the process is non-minimum phase. We shall study this issue, which is not absent in the tracking performance criterion either, in Sect. 2.4. It is also worthy of note the well known fact that if the system noise is white (that is, $H(z)=1$) then the minimum variance controller is $C_e(z)=0$—closing the loop can only worsen the noise rejection performance in this case.

If and only if the ideal controller $C_e(z)$ lies within the class of controllers considered, then the closed-loop system can be made to behave exactly as desired by a proper choice of the parameter ρ. Let us formalize this assumption.

Assumption B$_e$ $C_e(z)\in\mathcal{C}$ or, equivalently,

$$\exists\rho_e: C(z,\rho_e)=C_e(z). \tag{2.17}$$

Just like in the reference tracking case, whether or not this assumption is satisfied is critical in the study of the solution of the corresponding H_2 minimization problem. But Assumption B$_e$ tends to be more restrictive than Assumption B$_y$, because $T_e(z)$ is given, whereas $T_d(z)$ can be chosen—a consequence of not relaxing the noise rejection performance criterion.

Example 2.2 Consider the first-order system

$$G(z) = \frac{k}{(z-a)}$$
$$H(z) = \frac{z}{(z-b)}$$

with fixed but unknown $a,\ b \in (0,1), k \in \mathbb{R}$. The ideal controller regarding noise rejection is given by

$$C_e(z) = \frac{H(z)-1}{G(z)} = \frac{\frac{z}{(z-b)} - 1}{\frac{k}{(z-a)}}$$

which, after straightforward manipulation, simplifies to the following lead-lag controller

$$C_e(z) = \frac{b}{k}\frac{(z-a)}{(z-b)}.$$

2.3.3 The Composite Criterion

If the same controller happened to optimize the two different control objectives $J_y(\rho)$ and $J_e(\rho)$ at once, that is, if $C_d(z) = C_e(z)$, then the ideal controller for the composite performance criterion $J_T(\rho)$ would also be this same controller. This is, however, unlikely to happen in a real application, for a number of reasons; let us mention one of them. In most applications, one wants to track with zero steady-state error a reference with a given frequency, which requires a controller such that the loop gain tends to infinity at this frequency. In other words, the ideal controller in this case must satisfy:

$$C_d(e^{J\omega_1})G(e^{J\omega_1}) \to \infty \tag{2.18}$$

where ω_1 is the frequency of the reference to be tracked. If the two ideal controllers are the same, that is $C_d(z) = C_e(z)$, then (2.16) with (2.18) would imply that

$$H(e^{J\omega_1}) \to \infty \tag{2.19}$$

which violates one of the hypotheses on the noise, which is that the filter $H(z)$ is BIBO-stable. But this is not the only factor preventing the two controllers to be the same, as shown in the following example.

Example 2.3 Consider again the process in Example 2.2, with the following reference model

$$T_d(z) = \frac{1-c}{z-c}.$$

Then the ideal reference tracking controller is given by

$$
\begin{aligned}
C_d(z) &= \frac{\frac{1-c}{z-c}}{\frac{k}{(z-a)}(1-\frac{1-c}{z-c})} \\
&= \frac{\frac{1-c}{z-c}}{\frac{k}{(z-a)}(\frac{z-1}{z-c})} \\
&= \frac{(1-c)}{k}\frac{z-a}{z-1}
\end{aligned}
$$

which is a PI controller. Hence $C_e(z) = C_d(z)$ would require, according to (2.19) that $H(1) \to \infty$—the noise filter would have to be an integrator.

The ideal controller regarding noise rejection was calculated in Example 2.2 as

$$C_e(z) = \frac{b}{k}\frac{z-a}{z-b}$$

which equals $C_d(z)$ if and only if $b = 1$ and $c = 0$. That is, equality of the two ideal controllers would also require a specific reference model: $T_d(z) = \frac{1}{z}$.

So, the ideal reference tracking controller and the ideal noise rejection controller are hardly ever the same. This implies that the minimizer of the composite cost $J_T(\rho)$ depends not only on the process and noise characteristics—$G(z)$, $H(z)$—but also on the reference signal $r(t)$ and on the noise variance σ_e^2. This can be seen with a simple mind experiment. Assume that, in a particular problem, both ideal controllers belong to the controller class, that is $C_d(z), C_e(z) \in \mathcal{C}$. If the reference's amplitude is zero, then the composite cost equals the noise rejection cost—$J_T(\rho) = J_e(\rho)$—and the minimum of this criterion is achieved with $C_e(z)$, resulting in $J_T(\rho) = 0$. If, for the same process, the reference's amplitude is not zero, but the noise variance is, then the composite cost is given by the reference tracking alone—$J_T(\rho) = J_y(\rho)$—and its minimum becomes $C_d(z)$, again resulting in $J_T(\rho) = 0$. In the midway, as the signal to noise ratio becomes larger starting from zero, the optimal controller drifts away from $C_e(z)$, approaching $C_d(z)$ as the signal-to-noise ratio tends to infinity. Moreover, for any of these "midway" optimal controllers the resulting cost will not be zero—$J_T(\rho_*) \neq 0$. As a consequence, it is not possible for the composite performance criterion $J_T(\rho)$ to obtain a closed-form formula for an "ideal controller", as obtained in (2.9) and (2.16).

2.4 Beware of What You Ask for—Choosing the Reference Model

When applying a model reference design, the user is asking for the algorithm to find $C_d(z)$, which is the controller that provides the desired input-output relationship $\frac{y(z)}{r(z)} = T_d(z)$. In so doing, what the user has specified is "only" the input-output behavior of the closed-loop system, and the ideal controller, although it matches the desired input-output behavior, does not guarantee internal stability. While searching for his/her ideal controller, the user may find out that his/her ideal is not the best thing to do—maybe even disastrous. Besides, as we will see further ahead in this book, specifying a model reference that is too far from what can be achieved tends to complicate the optimization so that it becomes less likely that the best controller is ever achieved.

What can be done about it? Precautions can be taken in choosing the reference model to avoid such disastrous quest for an "ideal" that does not provide an appropriate behavior for the closed-loop system. This is what this section is about: these aspects are studied, and guidelines are derived for choosing the reference model safely, without waiving the desired performance. All this can be analyzed starting from $C_d(z)$, so we reproduce its expression here for ease of reference

$$C_d(z) = \frac{T_d(z)}{G(z)(1 - T_d(z))}. \tag{2.20}$$

In the choice of a sensible reference model, the first concern is that the ideal controller must be causal. Causality means that the relative degree of $C_d(z)$ is non-negative. Looking at (2.20) it is easy to verify that the relative degree of $C_d(z)$ equals the difference between the relative degree of the process and that of the reference model. Hence, in order to have a causal ideal controller, the reference model must be chosen according to the guideline below.

Guideline 1 (Causality of the controller) *The relative degree of the reference model $T_d(z)$ can not be smaller than the relative degree of the process $G(z)$.*

In order to be able to follow this guideline, it is necessary the *a priori* knowledge of an overbound for the relative degree of the process. This is the first instance found in this book of an obvious—yet sometimes forgotten—principle: it is impossible to properly design a controller for a given process without knowing anything at all about it. We are dealing with design methods that do not require the knowledge of the process' transfer function, but this does not mean that we do not know anything about the process. A process model is a rather complete and more or less exact description of the process' behavior for a wide class of excitations, and as such is often difficult and/or expensive to obtain. But the designer is likely to know basic features of the process with very little or no cost, such as an estimate of its static gain, whether or not it is stable and/or stably invertible, etc. So, what data-driven and adaptive design methods propose is not to design a controller completely in the dark, but instead to proceed with only some basic information about the process. Different methods will require different basic information.

Of course, one could just put an arbitrary controller in the loop, set randomly its parameters and hope that it will work; this does not require any knowledge about the process. Every now and then it will work—even a broken watch is correct at least once a day. But if we do not want to rely on luck to do our job, we are bound to use methods that are guaranteed to work well under reasonable circumstances, and such methods require knowledge of something about the process to be controlled. There does not exist a universal control/adaptation law, that would work properly for all processes in the world regardless of their static and dynamic properties and the performance requirements; searching for such a panacea is reminiscent of the quest for the *moto perpetuo*.

Let us close this brief philosophical parenthesis and go back to (2.20). There, we observe that the ideal controller is obtained through inversion of the process transfer function $G(z)$, so the transfer function of the ideal controller will have the zeros of the process as its poles and vice-versa. If any of the zeros or poles canceled by the ideal controller is outside the unit circle, then the resulting closed-loop with the ideal controller would be internally unstable, even though its input-output properties are the stable ones that were specified through the reference model. It is of course not acceptable to set the control design problem as the quest for a solution that is not even stabilizing. These unstable cancellations must be prevented at all costs and this can be accomplished by appropriately choosing the reference model. If the unstable singularities of the process are also present in the reference model, then they are canceled in the right hand side of (2.20) and do not appear in the ideal controller $C_d(z)$. We arrive thus at a second guideline.

Guideline 2 (Internal stability with non-minimum phase process) *The non-minimum phase zeros of the process $G(z)$ must be included in the reference model $T_d(z)$.*

Enforcing this second guideline requires the knowledge of the locations of the unstable zeros of the process, if any. Limitations in performance imposed by NMP zeros are well known and not including these zeros in the reference model would be fighting against nature. So, a successful model reference design requires the identification of the non-minimum phase zeros of the process. How to do this within the context of data-driven control design, without requiring an explicit previous identification procedure for the NMP zeros, is the subject of later discussion in Chap. 3.

2.4.1 Too Ambitious Performance

If the reference model $T_d(z)$ specifies a performance that is too different from the best that can be achieved within the given controller class, then the solution may have nothing to do with $T_d(z)$. Start with the following example.

Example 2.4 Consider a process described by the following transfer function

$$G(z) = \frac{z}{z - 0.2}$$

controlled by a purely integral controller

$$C(z, \rho) = \frac{\rho}{z - 1}.$$

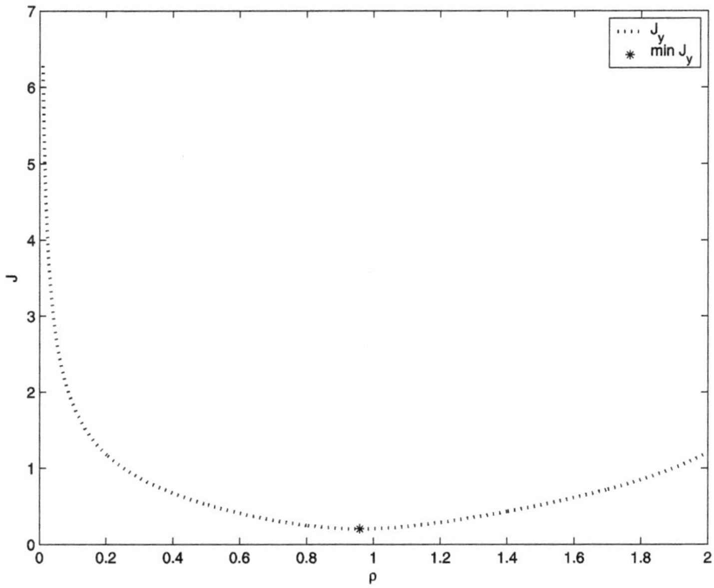

Fig. 2.2 Reference tracking performance criterion as a function of the scalar controller parameter ρ

The performance criterion is reference tracking, with the specification of the deadbeat reference model

$$T_{d1}(z) = \frac{1}{z}.$$

The corresponding ideal controller is given by

$$C_d(z) = \frac{z - 0.2}{z(z-1)}$$

which does not belong to the controller class considered. Then the minimum of the reference tracking criterion depends on the reference. For a step reference signal, the function $J_y(\rho)$ is as shown in Fig. 2.2, and it is observed that $\rho_* = 0.96$, that is, the optimal controller is $C(z, \rho_*) = \frac{0.96}{z-1}$. The transfer function of the closed-loop system with this controller becomes

$$T_1(z, \rho_*) = \frac{0.96z}{z^2 - 0.24z + 0.2}.$$

Now consider the alternative reference model

$$T_{d2}(z) = \frac{0.3z}{(z-0.5)(z-0.4)}$$

for which $C_d(z) = \frac{0.3}{z-1} \in \mathcal{C}$ is the optimal controller.

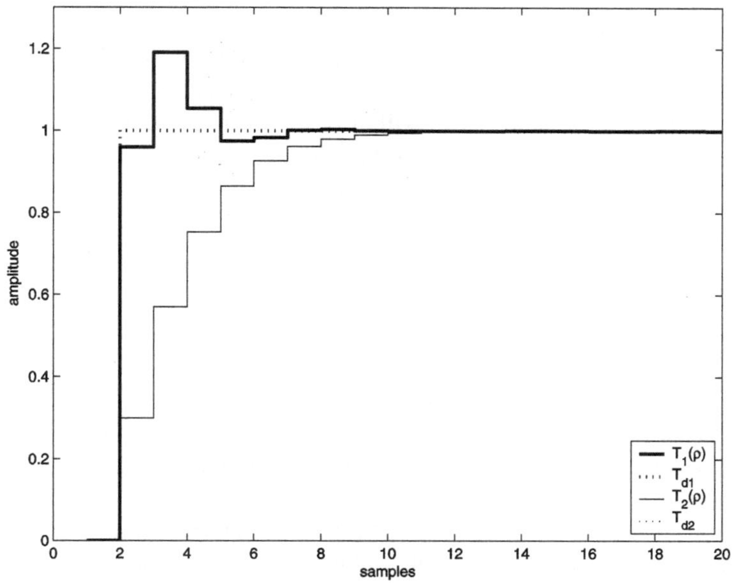

Fig. 2.3 System responses for a step applied at $t = 1$: deadbeat response specified by $T_{d1}(z)$ (dotted line), response achieved minimizing $J_y(\rho)$ with this reference model (thick line), and response specified by the reference model $T_{d2}(z)$, which is achieved exactly by minimizing $J_y(\rho)$ for this reference model

The step responses of the closed-loop system with each one of the optimal controllers found for each reference model are shown in Fig. 2.3. It is observed that the transient performance obtained with the problem formulation using the second reference model $T_{d2}(z)$ is better than the one obtained when the deadbeat reference model $T_{d1}(z)$ was used. This is because the closed-loop system can be made to behave exactly as specified by $T_{d2}(z)$ with the controller structure available, whereas for $T_{d1}(z)$ this is not possible.

To better appreciate the moral behind this example, recall that the original performance criterion that one would like to optimize is

$$\bar{E}[(T(z,\rho) - 1)r(t)]^2$$

which is $J_y(\rho)$ with $T_d(z) = 1$. The primary reason why a different $T_d(z)$ is specified instead of minimizing directly the cost function above is to make it possible to achieve the performance specified. So, it does not make a lot of sense to specify another reference model that is still far from what is achievable; if this were not a concern, we would just keep $T_d(z) = 1$.

Guideline 3 (Realistic ambition) *The reference model should be sufficiently close to what is possible to achieve with the given controller class.*

There is more than just intuition to support this guideline. Analytical results supporting this intuition will be presented in subsequent chapters showing that there exist a good number of properties of data-based methods that can only be proven when Assumption B_y is satisfied. It will also be shown that much can be done to improve these properties in this case, so that it is more likely and easier and safer to obtain optimality when this assumption is satisfied, or at least moderately violated—that is, when $\|K(z)\|$ is small.

2.5 Chapter Conclusions

This chapter presented a framework for the study of H_2 performance criteria, and all that has been said is inherent to this problem formulation. These properties apply to whatever control design based on this formulation. This includes data-driven control design methods, MRAC, and non-adaptive methods as well.

A central concept in this framework is the one of the ideal controller for each performance criterion—reference tracking and noise rejection. And central conditions are Assumptions B_y and B_e; whether or not these conditions are violated, and to which extent, is key to the properties of the data-driven design, as will be seen in the chapters to follow. A few principles and guidelines have also been established for the choice of the reference model, which must be followed in the problem setting. Verifying Assumptions B_y and/or B_e, and following these guidelines, are tasks that require some basic knowledge about the behavior of the process being controlled. There is no miracle: there is no need for models in data-driven control design, but the designer can not be completely in the dark; *some* (rudimentary) information about the process must be available.

Given the performance criterion, the design is nothing but an optimization problem, to be solved by some numerical algorithm. The H_2 control design thus consists "only" in choosing the reference model and then solving the optimization problem. Data-driven control design usually consists in performing the optimization based on data collected from the system, without the intermediate step of deriving a process model from these data. Data-driven control design methods rely mainly on iterative optimization procedures, mostly gradient descent algorithms. The quantities required in the optimization procedure are the cost function's gradient and possibly its Hessian, which are estimated pointwise directly from batches of input-output data collected from the closed-loop system. These details will be extensively analyzed along this book.

An alternative to iterative optimization methods is to approximate the cost function by a quadratic function whose minimum is the same. The function to be minimized being quadratic, no iterative algorithm is required, and the minimum can be found by one single least squares calculation. This alternative is presented in the next chapter.

References

1. K.J. Åström, B. Wittenmark, *Adaptive Control*, 2nd edn. (Addison-Wesley, Reading, 1995)
2. M.C. Campi, A. Lecchini, S.M. Savaresi, Virtual reference feedback tuning: A direct method for the design of feedback controllers. Automatica **38**, 1337–1346 (2002)
3. G. Goodwin, S.F. Graebe, M.E. Salgado, *Control System Design* (Prentice Hall, New York, 2000)
4. L. Ljung, *System Identification – Theory for the User*, 2nd edn. (Prentice Hall, New York, 1999)

Chapter 3
One-Shot Optimization—The VRFT Method

3.1 Introduction

Virtual Reference Feedback Tuning (VRFT) is a method for optimizing the reference tracking criterion $J_y(\rho)$ whose most attractive feature is the fact that it is not iterative. The core idea of this method is to cast the design problem as the Prediction Error Identification of a model for the ideal controller $C_d(z)$ in Auto-Regressive form. Prediction Error Identification of such a model consists in minimizing a quadratic function of the model parameters, which we shall refer to as $J^{VR}(\rho)$. This is a function whose global minimum is, under ideal conditions, the same as the global minimum of $J_y(\rho)$, but whose minimization is much less demanding because it is a quadratic function. Hypothetical cost functions $J_y(\rho)$ and $J^{VR}(\rho)$ for a scalar parameter ρ are depicted in Fig. 3.1 to illustrate this situation. Optimization of the original reference tracking cost function $J_y(\rho)$ can prove troublesome due to the local extrema observed in this function, whereas $J^{VR}(\rho)$ does not suffer from this inconvenient.

In this chapter we will present the VRFT method in detail, starting with the ideal case, for which the method is originally conceived and in which the concepts appear more clearly.

3.2 The Ideal Case

The Virtual Reference Feedback Tuning method is conceived with an ideal situation in mind, in which:

- the system is not affected by noise, that is, $\sigma_e = 0$;
- the ideal controller (2.9) belongs to the considered controller class, that is, Assumption B_y is satisfied;
- the controller is parameterized linearly, that is, Assumption LP is satisfied.

From this idealized situation the main ideas are derived, and extensions for the nonideal case are developed later. The VRFT method can be described as follows.

A. Sanfelice Bazanella et al., *Data-Driven Controller Design*,
Communications and Control Engineering,
DOI 10.1007/978-94-007-2300-9_3, © Springer Science+Business Media B.V. 2012

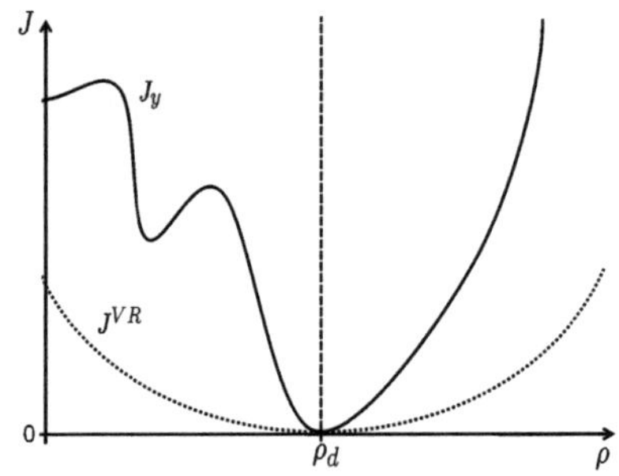

Fig. 3.1 Illustrative plots of the cost functions $J_y(\rho)$ and $J^{VR}(\rho)$ for the case of a scalar parameter

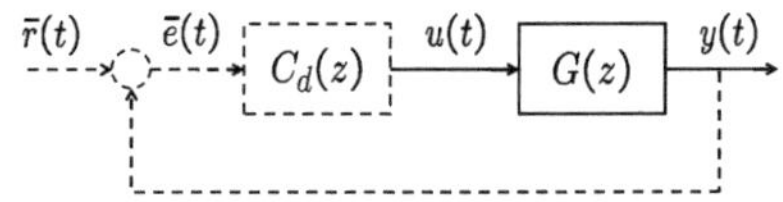

Fig. 3.2 Experiment for obtaining data for the VRFT method: real data (*solid lines*) and virtual data (*dashed lines*)

Through either an open-loop or a closed-loop experiment, input data $u(t)$ and output data $y(t)$ are collected from the process. Given the measured $y(t)$, we define the *virtual reference* signal $\bar{r}(t)$ such that

$$T_d(z)\bar{r}(t) = y(t),$$

where $T_d(z)$ is the desired reference model for the closed-loop response.

If the system were operating in closed loop with the ideal controller $C_d(z)$, and the virtual reference $\bar{r}(t)$ were applied, the output data would have been the same as the data $y(t)$ and $u(t)$ that have actually been collected. The VRFT method goes on as if this fake experiment had actually been performed. Should the data have been collected like this, the reference tracking error would have been given by

$$\bar{e}(t) = \bar{r}(t) - y(t).$$

This $\bar{e}(t)$ is the signal that would have fed the ideal controller in this virtual experiment, as presented in Fig. 3.2. We thus have input and output data ($\bar{e}(t)$ and $u(t)$ respectively) from the ideal controller $C_d(z)$ in the fake experiment and we can use these data to identify it. With $\bar{e}(t)$ being the controller's input, $C(z, \rho)$ a model for the controller, and $C(z, \rho)\bar{e}(t)$ the predicted controller's output with this model, a Prediction Error Identification criterion is formed as the H_2 norm of the prediction error; this is the VRFT criterion $J^{VR}(\rho)$ previously mentioned:

$$J^{VR}(\rho) \triangleq \bar{E}\left[u(t) - C(z, \rho)\bar{e}(t)\right]^2. \tag{3.1}$$

For a linearly parameterized controller as in Assumption LP, the criterion $J^{VR}(\rho)$ in (3.1) can be written as

$$\begin{aligned} J^{VR}(\rho) &= \bar{E}[u(t) - \rho^T \bar{C}(z)\bar{e}(t)]^2 \\ &= \bar{E}[u(t) - \rho^T \varphi(t)]^2, \end{aligned} \tag{3.2}$$

where the *regressor vector* $\varphi(t)$ is defined as

$$\varphi(t) = \bar{C}(z)\bar{e}(t) = \bar{C}(z)\frac{1 - T_d(z)}{T_d(z)}y(t).$$

This is a standard formulation in system's parameter identification, where the system to be identified is the ideal controller, $u(t)$ is its measured output and $\bar{e}(t)$ its measured input. By noting that the data are generated by

$$u(t) = C_d(z)\bar{e}(t)$$

the prediction error identification cost function $J^{VR}(\rho)$ can be further developed into

$$\begin{aligned} J^{VR}(\rho) &= \bar{E}[(\rho_d - \rho^T)\varphi(t)]^2 \\ &= \bar{E}\left[(\rho_d - \rho^T)\bar{C}(z)\frac{1 - T_d(z)}{T_d(z)}G(z)u(t)\right]^2, \end{aligned} \tag{3.3}$$

or, alternatively,

$$J^{VR}(\rho) = \bar{E}\left[(\rho_d - \rho^T)\bar{C}(z)\frac{1}{C_d(z)}u(t)\right]^2. \tag{3.4}$$

Application of Parseval's theorem to (3.4) yields

$$J^{VR}(\rho) = \frac{1}{2\pi}\int_{-\pi}^{\pi}\frac{1}{|C_d(e^{j\omega})|^2}|(\rho - \rho_d)^T\bar{C}(e^{j\omega})|^2\Phi_u(e^{j\omega})d\omega \tag{3.5}$$

provided that either $C_d(z)$ has no zeros on the unit circle or these zeros do not correspond to frequencies in the support of $u(t)$; otherwise the integral does not exist. It is clear in (3.5) that $J^{VR}(\rho_d) = 0$ and therefore ρ_d is a global minimum of $J^{VR}(\rho)$. Moreover, (3.5) can be reorganized as a quadratic form:

$$J^{VR}(\rho) = (\rho - \rho_d)^T A^{VR}(\rho - \rho_d), \tag{3.6}$$

where

$$A^{VR} = \frac{1}{2\pi}\int_{-\pi}^{\pi}\frac{1}{|C_d(e^{j\omega})|^2}\bar{C}(e^{j\omega})\bar{C}^*(e^{j\omega})\Phi_u(e^{j\omega})d\omega \tag{3.7}$$

is a positive semi-definite matrix by construction. It is clear that ρ_d is a global minimum of this function, as desired, and that it is the unique global minimum provided that A^{VR} is a positive definite matrix. Let us assume that the input has a discrete spectrum, that is

$$\Phi_u(e^{j\omega}) = \sum_{k=1}^{q}\lambda_k\delta(\omega - \omega_k)$$

where $\delta(\cdot)$ is the Dirac's delta function and λ_k are positive real numbers. Then

$$A^{VR} = \frac{1}{2\pi} \sum_{k=1}^{q} \frac{1}{|C_d(e^{J\omega_k})|^2} \bar{C}(e^{J\omega_k})\bar{C}^*(e^{J\omega_k})\lambda_k. \tag{3.8}$$

Each one of the matrices $\bar{C}(e^{J\omega_k})\bar{C}^*(e^{J\omega_k})$ is a positive semi-definite matrix, and the matrix A^{VR} is a weighed sum of q of these matrices. The rank of such a sum is $\min(q, p)$ as a generic property, that is, for almost all spectra with q nonzero components. This is a standard result in adaptive control and in identification theory whose demonstration will not be detailed here. The curious reader (and the skeptical one) can check the formal statement of this and the following properties in the literature; some references for this purpose are given in this chapter's conclusions.

We define the following nomenclature, which is classic in the identification literature and will be used extensively in this book.

Definition 3.1 A quasi-stationary process is said to be sufficiently rich of order q—or SRq for short—if its spectrum has at least q nonzero components.

With this definition, we can say that the optimal controller parameter ρ_d is the unique global minimum of $J^{VR}(\rho)$ if the input signal is sufficiently rich of order p.

Coming back to the standard equation (3.2), the problem of finding the global minimum of a function of this form is known in the literature as the Least Squares problem, whose solution is given by the solutions ρ_* of the following system of linear equations, known as the *normal equation*:

$$\bar{E}[\varphi(t)\varphi^T(t)]\rho_* = \bar{E}\left[\varphi(t)u(t)\right]. \tag{3.9}$$

The solution of the normal equation exists and is unique if the matrix $\bar{E}[\varphi(t) \times \varphi^T(t)]$ is full rank. This property of the regressor vector is given a name in the literature.

Definition 3.2 A quasi-stationary vector $\varphi(t)$ is said to be persistently exciting if $\bar{E}[\varphi(t)\varphi^T(t)] > 0$.

The regressor vector is generated by filtering of the input $u(t)$, so the persistent excitation condition, which is needed for the solution of the normal equation to be unique, can be translated into conditions on $u(t)$. This is the so-called problem of "transfer of excitation", and it turns out that, apart some technicalities, one can say that the regressor vector is persistently exciting if $u(t)$ is SRp.[1] So, we have arrived by two different ways at the same conclusion: under Assumptions B_y and LP, and with the SRp condition, the unique global minimum of $J^{VR}(\rho)$ is also the global minimum of the reference tracking criterion $J_y(\rho)$. Minimizing $J^{VR}(\rho)$ under this

[1] Again, we leave to the reader to check the details in the literature.

Assumption yields the desired result, which is the global minimum of the reference tracking performance criterion $J_y(\rho)$. This is the whole point of the VRFT method.

But (3.9) is not a directly computable quantity, since it involves expectations. Given a set of N input-output data collected from the system, the solution ρ_* of the normal equation is obtained through the following calculation:

$$\hat{\rho} = \left[\sum_{t=1}^{N} \varphi(t)\varphi^T(t) \right]^{-1} \sum_{t=1}^{N} \varphi(t)u(t) = \rho_*. \tag{3.10}$$

This calculation will result exactly in the asymptotic solution ρ_* only in this ideal case, where there is no noise. When there is noise in the system, then the parameter $\hat{\rho}$ resulting from (3.10) is a random variable, whose properties will be discussed a few pages ahead. In any case, this "one-shot" (that is, noniterative) feature is the key advantage of working with the VRFT criterion (3.1) over the model reference criterion $J_y(\rho)$.

In summary, considering that the signals measured are free of noise and that the input is SRp, and under Assumptions LP and B_y, the VRFT procedure identifies the ideal controller (2.9) exactly. We illustrate the calculations involved in the VRFT method in an example.

Example 3.1 Let the open-loop system be given by $y(t) = G(z)u(t)$ with

$$G(z) = \frac{0.5}{z - 0.9}$$

and the reference model be of the form

$$T_d(z) = \frac{1-a}{z-a},$$

with the parameter $a \in (0, 1)$ to be specified. Notice that $T_d(1) = 1$, so that zero steady-state error for a step reference is obtained, regardless of the value of a.

In order to be in the ideal case, we need to choose the controller class so that the ideal controller belongs to it. Knowing that the process can be modeled by a first order transfer function

$$G(z) = \frac{b_1}{z - b_2},$$

the ideal controller is given by

$$\begin{aligned} C_d(z) &= \frac{\frac{1-a}{z-a}}{\frac{b_1}{z-b_2}(1 - \frac{1-a}{z-a})} \\ &= \frac{(1-a)(z-b_2)(z-a)}{(z-a)[b_1(z-a) - b_1(1-a)]} \\ &= \frac{1-a}{b_1} \frac{z-b_2}{z-1}, \end{aligned} \tag{3.11}$$

Table 3.1 Signals used to minimize $J^{VR}(\rho)$ in the matched case example

$u(t)$	$y(t)$	$\bar{r}(t)$	$\bar{e}(t)$	$\varphi(t)$
1.000	0.000	1.250	1.250	$\begin{bmatrix} 1.250 & 0.000 \\ 2.375 & 1.250 \end{bmatrix}$
1.000	0.500	1.625	1.125	

which is a PI controller. We then choose the controller class to be the class of PI controllers, that is

$$\bar{C}(z) = \begin{bmatrix} \frac{z}{z-1} \\ \frac{1}{z-1} \end{bmatrix};$$

in so doing, Assumption B_y is satisfied.

Now let us specify the parameter $a = 0.6$ which amounts to the reference model

$$T_d(z) = \frac{0.4}{z - 0.6}.$$

We apply a step of amplitude 1 as the input signal of an open-loop experiment and collect only $N = 2$ samples[2] of the input signal $u(t)$ and the output signal $y(t)$. Using $\bar{C}(z)$ and $T_d(z)$ and the collected data, we can calculate the signals presented in Table 3.1; we then minimize $J^{VR}(\rho)$ (3.1) through (3.10) and obtain

$$\hat{\rho} = \begin{bmatrix} 0.8 \\ 0.72 \end{bmatrix}.$$

Substituting $a = 0.6$, $b_1 = 0.5$ and $b_2 = 0.9$ into (3.11) the ideal controller is found to be

$$C_d(z) = 0.8\frac{z - 0.9}{z - 1} = [0.8 \;\; -0.72] \begin{bmatrix} \frac{z}{z-1} \\ \frac{1}{z-1} \end{bmatrix},$$

exactly the same we have found using the VRFT method.

3.2.1 Generation of the Virtual Reference—A Caveat

The scenario where Assumption B_y is satisfied can usually be forged by the user, which is done by choosing appropriately the controller class and the reference model. This must be done without knowing the process model, but only some of its features, as explained in Chap. 2. As discussed in Sect. 2.4, the choice of $T_d(z)$ should be guided by the process constraints in order for the algorithm be able to find $C_d(z)$, that is, to satisfy Assumption B_y. Specifically, when the process is

[2]Since we are estimating a PI controller, which is formed by a parameter vector of size two, and we are dealing with a noise-free case, we can use the minimum amount of data.

non-minimum phase, then its non-minimum zeros should be zeros of the reference model. This brings about an implementation issue for the VRFT method, since the virtual reference is calculated with the inverse of $T_d(z)$, which in this case would be unstable.

This issue can be dealt with by multiplying the signals $u(t)$ and $y(t)$ with an all-pass frequency weighting filter, which leaves the objective function $J^{VR}(\rho)$ (3.1) unchanged; the signals needed by the VRFT method are then obtained from stable filters [9, 16]. We briefly explain how this all-pass filter is obtained.

Let $T_d(z)$ be factored as

$$T_d(z) = \frac{n_u(z)n_s(z)}{d(z)},$$

where the factor n_u

$$n_u(z) = \prod_{k=1}^{m}(z - x_k)$$

contains all the unstable zeros and nothing more. Let $L_{ap}(z)$ be the following all-pass filter

$$L_{ap}(z) \triangleq \frac{n_u(z)}{n_u^*(z)}$$

that is, a filter whose magnitude is 1 for all frequencies, where

$$n_u^*(z) = \prod_{k=1}^{m}(x_k z - 1).$$

Thus, the criterion $J^{VR}(\rho)$ is slightly changed in such a way that a stable filter is used to calculate the virtual reference signal. Let

$$T_a(z) = \frac{n_u^*(z)n_s(z)}{d(z)}$$

be a transfer function whose magnitude is the same as the magnitude of $T_d(z)$, but with stable zeros only. Then, using $L_{ap}(z)$, we have that

$$J_a^{VR}(\rho) = \bar{E}\left\{L_{ap}(z)\left[u(t) - \left(C(z,\rho)\frac{1 - T_a(z)}{T_a(z)}\right)y(t)\right]\right\}^2$$

whose minimum is the same as the minimum of $J^{VR}(\rho)$. This is the cost to be minimized in order to find the controller parameters when the reference model contains NMP zeros.

Another possibility is to multiply the $J^{VR}(\rho)$ criterion by $T_d(z)$, so that (3.1) becomes

$$J_{a'}^{VR}(\rho) \triangleq \bar{E}[T_d(z)u(t) - (C(z,\rho)(1 - T_d(z)))y(t)]^2 \qquad (3.12)$$

whose minimum is the same as the minimum of the VRFT criterion, but we do not have the inconvenience of unstable filtering of the signals. The same artifice is adopted in Sect. 3.4.

The ideal case, as in the example presented above, is rarely found. In many cases, Assumption B_y can be made to be satisfied within any given precision by a proper choice of the reference model, but this is not always possible. Moreover, the noise levels are often small, or can be made so by appropriate filtering, but again this is not generic. And any violation of the basic hypothesis will result in a solution that is not the one desired, for the minima of $J^{VR}(\rho)$ and $J_y(\rho)$ only coincide under these ideal conditions. In most real applications, the VRFT method needs some modifications to deal with these real-life facts: that it is impossible to achieve exactly the ideal controller and that the signals are corrupted by noise. Let us now show the techniques that are used to handle these issues, starting with the violation of Assumption B_y.

3.3 The Mismatched Case

When Assumption B_y is not satisfied, that is, $C_d(z) \notin \mathcal{C}$, we say that the control is mismatched. In the matched case, $\arg\min J^{VR}(\rho) = \arg\min J_y(\rho)$, but this will no longer be the case when Assumption B_y is not satisfied.

Let us compare the two performance criteria by looking at their frequency domain expressions. Applying Parseval's Theorem to the reference tracking cost function

$$J_y(\rho) = \bar{E}[(T(z,\rho) - T_d(z))r(t)]^2$$

and using the closed-loop relations

$$T_d(z) = \frac{C_d(z)G(z)}{1 + C_d(z)G(z)} \qquad T(z,\rho) = \frac{C(z,\rho)G(z)}{1 + C(z,\rho)G(z)}$$

results in the following frequency domain expression for the cost function:

$$J_y(\rho) = \frac{1}{2\pi}\int_{-\pi}^{\pi} \left| \frac{G(e^{j\omega})C(e^{j\omega},\rho)}{1 + G(e^{j\omega})C(e^{j\omega},\rho)} - \frac{G(e^{j\omega})C_d(e^{j\omega})}{1 + G(e^{j\omega})C_d(e^{j\omega})} \right|^2 \Phi_r(e^{j\omega})d\omega.$$

Developing this expression using the least common denominator yields

$$\begin{aligned} J_y(\rho) &= \frac{1}{2\pi}\int_{-\pi}^{\pi} \left| \frac{G(e^{j\omega})[C(e^{j\omega},\rho)(1 + G(e^{j\omega})C_d(e^{j\omega})) - C_d(e^{j\omega})(1 + G(e^{j\omega})C(e^{j\omega},\rho))]}{[1 + G(e^{j\omega})C(e^{j\omega},\rho)][1 + G(e^{j\omega})C_d(e^{j\omega})]} \right|^2 \\ &\quad \times \Phi_r(e^{j\omega})d\omega \\ &= \frac{1}{2\pi}\int_{-\pi}^{\pi} \frac{|G(e^{j\omega})|^2 |C(e^{j\omega},\rho) - C_d(e^{j\omega})|^2}{|1 + G(e^{j\omega})C(e^{j\omega},\rho)|^2 |1 + G(e^{j\omega})C_d(e^{j\omega})|^2} \Phi_r(e^{j\omega})d\omega \end{aligned}$$

which is written more compactly as

$$J_y(\rho) = \frac{1}{2\pi}\int_{-\pi}^{\pi} |G(e^{J\omega})|^2 |S(e^{J\omega},\rho)|^2 |S_d(e^{J\omega})|^2 \\ \times |C(e^{J\omega},\rho) - C_d(e^{J\omega})|^2 \Phi_r(e^{J\omega}) d\omega. \tag{3.13}$$

As for the VRFT objective function $J^{VR}(\rho)$, let us first include a filtering $L(z)$ to the signals $u(t)$ and $\bar{e}(t)$, so that it becomes

$$\begin{aligned} J^{VR}(\rho) &= \bar{E}\,[L(z)(u(t) - C(z,\rho)\bar{e}(t))]^2 \\ &= \bar{E}\left[L(z)\left(u(t) - \left(C(z,\rho)\frac{1-T_d(z)}{T_d(z)}\right)y(t)\right)\right]^2. \end{aligned} \tag{3.14}$$

The role of this filter will be seen in a moment. Now, applying Parseval's theorem to (3.14) yields

$$\begin{aligned} J^{VR}(\rho) &= \frac{1}{2\pi}\int_{-\pi}^{\pi} \left| L(e^{J\omega})\left[1 - \left(C(e^{J\omega},\rho)\frac{1-T_d(e^{J\omega})}{T_d(e^{J\omega})}\right)G(e^{J\omega})\right]\right|^2 \Phi_u(e^{J\omega})d\omega \\ &= \frac{1}{2\pi}\int_{-\pi}^{\pi} |L(e^{J\omega})|^2 \frac{|G(e^{J\omega})|^2}{|T_d(e^{J\omega})|^2} \\ &\quad \times \left|\frac{T_d(e^{J\omega})}{G(e^{J\omega})} - C(e^{J\omega},\rho)(1 - T_d(e^{J\omega}))\right|^2 \Phi_u(e^{J\omega})d\omega. \end{aligned}$$

But

$$1 - T_d(e^{J\omega}) = S_d(e^{J\omega})$$

and

$$\frac{T_d(e^{J\omega})}{G(e^{J\omega})} = C_d(e^{J\omega})S_d(e^{J\omega}),$$

so we can finally write $J^{VR}(\rho)$ as

$$J^{VR}(\rho) = \frac{1}{2\pi}\int_{-\pi}^{\pi} |L(e^{J\omega})|^2 \frac{|G(e^{J\omega})|^2 |S_d(e^{J\omega})|^2}{|T_d(e^{J\omega})|^2} \\ \times |C_d(e^{J\omega}) - C(e^{J\omega},\rho)|^2 \Phi_u(e^{J\omega})d\omega. \tag{3.15}$$

Compare now (3.13) with (3.15). When Assumption B_y is satisfied both cost functions have a global minimum at the same parameter value ρ_d (this does not even depend on the linear parameterization hypothesis LP). But when Assumption B_y is not satisfied, neither criterion ever vanishes and the minima of either one depend on the various factors inside the integral multiplying the difference $|C_d(e^{J\omega}) - C(e^{J\omega},\rho)|^2$. If these factors in the two integrands are different, there is no reason to expect that the two minima will be the same.

But there is a free parameter that has been included in the VRFT design and that can be chosen such that these two integrands are alike: the filter $L(z)$. In order to make $J^{VR}(\rho) = J_y(\rho)$, it suffices to chose $L(e^{J\omega})$ as follows:

$$|L(e^{J\omega})|^2 = |T_d(e^{J\omega})|^2 |S(e^{J\omega}, \rho)|^2 \frac{\Phi_r(e^{J\omega})}{\Phi_u(e^{J\omega})}, \quad \forall \omega \in [-\pi; \pi], \tag{3.16}$$

because then both integrands will be the same. Notice that $\Phi_r(e^{J\omega})$ here comes from the criterion $J_y(\rho)$, hence it represents the spectrum of the real reference signal $r(t)$ which will be applied to the system during its operation, not the virtual reference. And that $\Phi_u(e^{J\omega})$ comes from the VRFT criterion $J^{VR}(\rho)$, so it represents the spectrum of the input $u(t)$ actually measured in the VRFT experiment.

If a filter $L(z)$ whose magnitude satisfies (3.16) is applied to the data in the VRFT procedure, then both cost functions are the same and hence obviously both minima will be the same. But there is a catch, of course: the calculation of the transfer function desired for the filter $L(z)$ as specified in (3.16) requires the knowledge of $S(z, \rho)$, something that is not available. Then the implementation of this filter will rely on some approximation of this transfer function. The better this approximation, the closer the minimum that will be calculated by VRFT (the minimum of $J^{VR}(\rho)$) will be to the desired result, which is the minimum of $J_y(\rho)$.

Although this is probably not the only sensible choice for an approximation of $S(z, \rho)$, VRFT invariably uses the following:

$$|S(e^{J\omega}, \rho)|^2 \approx |S_d(e^{J\omega})|^2 = |1 - T_d(e^{J\omega})|^2, \tag{3.17}$$

which appears to be a sensible approximation, since we expect the two sensibilities in (3.17) to be very close to each other around the minimum. Using this approximation, the appropriate filter's transfer function can be obtained from

$$|L(e^{J\omega})|^2 = |T_d(e^{J\omega})|^2 |1 - T_d(e^{J\omega})|^2 \frac{\Phi_r(e^{J\omega})}{\Phi_u(e^{J\omega})}, \quad \forall \omega \in [-\pi; \pi], \tag{3.18}$$

since all quantities in the right side are known. In fact, $\Phi_u(e^{J\omega})$ can be considered known only when the input signal has been selected by the designer; in other situations, it must be estimated. Actually, if the user can choose the input signal of an open loop experiment to be the same type as the reference signal which is usually applied on the process, then $\frac{\Phi_r(e^{J\omega})}{\Phi_u(e^{J\omega})} = 1$, and the filter is only dependent on $T_d(z)$, which is known.

Thus, in the mismatched case, the filter is computed using (3.18) and the asymptotic value of the parameter vector ρ is given by

$$\rho_* = \bar{E}[\varphi_L(t)\varphi_L(t)^T]^{-1} \bar{E}[\varphi_L(t)u_L(t)] \tag{3.19}$$

where $\varphi_L(t) = L(z)\varphi(t)$ and $u_L(t) = L(z)u(t)$. The actual calculation used to obtain the optimal parameter value is similar to (3.10):

$$\hat{\rho} = \left[\sum_{t=1}^{N} \varphi_L(t)\varphi_L^T(t)\right]^{-1} \sum_{t=1}^{N} \varphi_L(t)u_L(t),$$

and in the absence of noise $\hat{\rho}$ equals the asymptotic value ρ_* in (3.19).

Example 3.2 Let us apply the VRFT method to find the best reference tracking controller for a given process whose transfer function is given by

$$G(z) = \frac{0.5(z - 0.5)}{(z - 0.9)^2}.$$

The VRFT design will not use the knowledge of the process' transfer function to proceed, of course. This system must track a step reference. Suppose that only an integrator is available to control the process, that is, the controller is given by

$$C(z) = \rho \frac{z}{z - 1}$$

and that the following reference model is chosen:

$$T_d(z) = \frac{1 - a}{z - a}, \tag{3.20}$$

where a can be chosen arbitrarily. The fact that the real process is a second order process, together with the controller class given and the reference model chosen, characterizes the mismatched case. Thus, the use of the filter $L(z)$ is important to approximate the minimum of the VRFT to the actual design objective, that is, the minimum of $J_y(\rho)$.

Consider the application of a VRFT design with $N = 500$ data samples collected in an experiment consisting of the application of a unit step as an open-loop input signal $u(t)$. Since this signal is the same as the reference that the system is supposed to track, the filter $L(z)$ is obtained through (3.18) using $\frac{\Phi_r(e^{j\omega})}{\Phi_u(e^{j\omega})} = 1$. In order to see the effectiveness of the filter, we compare the plots of the objective function $J_y(\rho)$, the VRFT criterion $J^{VR}(\rho)$ without filter (3.1), and $J^{VR}(\rho)$ with the filter $L(z)$ (3.14). We show the cost functions and the respective minimum values ρ_* for reference models obtained using different values of a.

Figure 3.3 shows the cost functions' plots for the reference model (3.20) with $a = 0.7$.[3] The minimum values of the costs and their corresponding arguments are presented in Table 3.2. Note that with the use of the filter the minimum of the VRFT criterion gets much closer to the minimum of the model reference (MR) criterion $J_y(\rho)$. Probably even more important, the value of the reference tracking objective function $J_y(\rho_*)$ decreases significantly, from 12.0802 to 5.36806. Compare these values with the value at the global minimum 3.71748: the VRFT with the filter yields a 44% surplus in cost with respect to the global minimum, whereas the design with no filter results in a cost that is over three times larger than at the global minimum.

The values presented in Table 3.3 are obtained for a reference model (3.20) with $a = 0.9$. The use of the filter again causes the minimum of the VRFT cost to approach the minimum of $J_y(\rho)$; however, for this new value of a, which represents a

[3] The introduction of the filter causes the cost function $J^{VR}(\rho)$ to become flatter, so we have plotted $200J^{VR}(\rho)$ to make the visual comparison to the other costs easier.

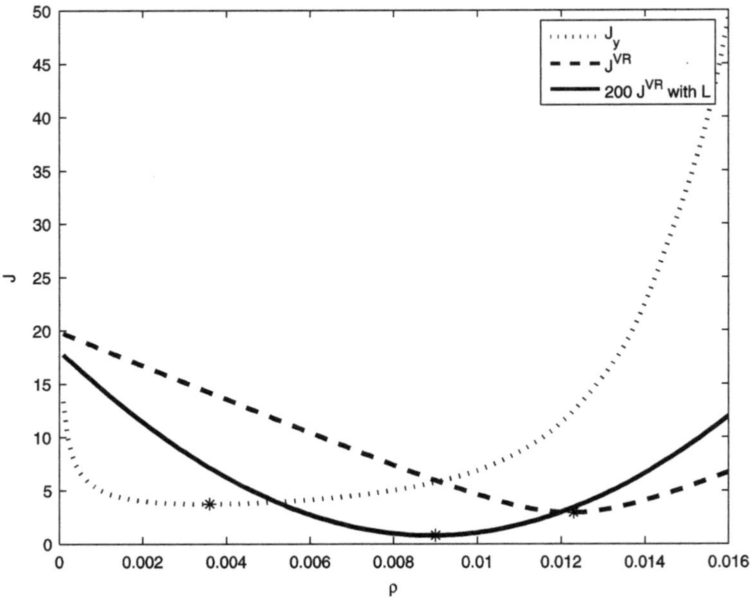

Fig. 3.3 Plots of $J_y(\rho)$ and $J^{VR}(\rho)$ with and without using the filter $L(z)$, varying the parameter ρ for a reference model (3.20) with $a = 0.7$; asterisks denote the minimum of each cost

Table 3.2 Minimizers of $J_y(\rho)$ and $J^{VR}(\rho)$ with and without using the filter $L(z)$ and their relative cost values for a reference model (3.20) with $a = 0.7$

Method	ρ_*	$J_y(\rho_*)$
MR	0.00350	3.71748
VRFT	0.01220	12.0802
VRFT + $L(z)$	0.00850	5.36806

Table 3.3 Minimizers of $J_y(\rho)$ and $J^{VR}(\rho)$ with and without using the filter $L(z)$ and their relative cost values for a reference model (3.20) with $a = 0.9$

Method	ρ_*	$J_y(\rho_*)$
MR	0.00290	2.69770
VRFT	0.00410	2.86835
VRFT + $L(z)$	0.00350	2.74450

slower reference model, both the value of the minimum ρ_* and the minimum value of the cost $J_y(\rho_*)$ are much closer to the values of the reference tracking criterion than in the case $a = 0.7$. This is because with this reference model the violation of Assumption B_y is less important, that is, $\|C(z, \rho_*) - C_d(z)\|$ is smaller.

On the other hand, a disastrous result is obtained if a much faster reference model were specified, for example $a = 0.3$. The minimizers and the minimum cost values for this case are presented in Table 3.4. Differently from the cases above, now the VRFT method results in a controller that makes the closed-loop unstable, regardless

Table 3.4 Minimizers of $J_y(\rho)$ and $J^{VR}(\rho)$ with and without using the filter $L(z)$ and their relative cost values for a reference model (3.20) with $a = 0.3$

Method	ρ_*	$J_y(\rho_*)$
MR	0.00400	4.09444
VRFT	0.02900	5496.90
VRFT + $L(z)$	0.01700	75.25891

the use of the filter. The closed-loop transfer functions, obtained with and without the filter $L(z)$, are given by

$$T(z, 0.02900) = \frac{0.0145z(z - 0.5000)}{(z - 0.7800)(z^2 - 2.005z + 1.038)} \tag{3.21}$$

and

$$T(z, 0.01700) = \frac{0.0085z(z - 0.5000)}{(z - 0.7993)(z^2 - 1.992z + 1.013)} \tag{3.22}$$

respectively, both of which present unstable poles.

When Assumption B_y is satisfied VRFT provides exactly the desired result, which is the minimum of $J_y(\rho)$. The example above illustrates that for moderate violations of Assumption B_y, that is, when the performance specified by the reference model can "almost" be exactly achieved, then the incorporation of the proper filter in the VRFT method succeeds in providing almost the same result. As the violation of this assumption becomes more pronounced, the filter still helps, but gradually loses its effectiveness. But if Assumption B_y is really strongly violated, that is, if the best performance that can be achieved with the controller class available is really far from the reference model, then the VRFT method becomes ineffective, regardless of the use of the filter.

This is expected, since the formulation of the filter $L(z)$ (3.18) is based on the approximation (3.17). If this approximation is not valid, the closed-loop response obtained from VRFT may be far from the desired one and, in some cases, even unstable. This undesired situation is also observed when the VRFT method is applied to non-minimum phase processes. This special case is dealt with in the next section, where we show a modification in the criterion so that the VRFT method can also be safely applied to this class of processes.

3.4 Dealing with Non-minimum Phase Plants

The VRFT method searches the parameters of a fixed structure controller to make the behavior of the closed-loop system as close as possible to the closed-loop system with $C_d(z)$. When the process has non-minimum phase (NMP) zeros and the reference model does not, the ideal controller $C_d(z)$ cancels out the NMP zeros of the process. Therefore the closed-loop system with $C_d(z)$ would be internally unstable and VRFT would try to mimic this disastrous behavior. One could hope that if Assumption B_y is not satisfied then the controller will not be able to mimic the unstable behavior and disaster will be avoided. The following example illustrates that this hope is not justified, and thus including the NMP zeros in the reference model is a safeguard to be applied in all cases.

Example 3.3 Let the process to be controlled be a non-minimum phase process, described by

$$G(z) = \frac{(1.1 - z)(z - 0.8)}{(z - 0.9)^3}$$

and the reference model be

$$T_d(z) = \frac{0.125z^2}{(z - 0.5)^3}.$$

In this case, the ideal controller would be given by

$$C_d(z) = \frac{-0.125z^2(z - 0.9)^3}{(z - 1.1)(z - 1)(z - 0.8)(z^2 - 0.625z + 0.125)}$$

whose application would cause internal instability. We would like to design a PID controller applying the VRFT method, with a transfer function given by (1.4). This transfer function has fixed poles, so the NMP zeros of the plant will not be canceled by the controller. The design is performed for the linear parameterization presented in (1.4), that is, the gains k_p, k_i and k_d are calculated, but the resulting controller is presented in the zero-pole form to facilitate the stability analysis.

Suppose that there is no noise in the output of the process and that 600 samples of the data are collected from an open-loop experiment, where

$$u(t) = 0.5 + 0.5sq\left(\frac{2\pi}{200}t\right),$$

where $sq(\frac{2\pi}{T}t)$ denotes a square wave with period T. The application of VRFT to design a PID controller results in the following controller:

$$C(z, \rho) = \frac{0.2764(z - 1.176)(z - 0.957)}{z(z - 1)}$$

and the corresponding closed-loop transfer function is

$$T(z, \rho) = \frac{-0.2764(z - 1.176)(z - 1.1)(z - 0.957)(z - 0.8)}{(z - 1.024)(z - 0.9523)(z - 0.4684)(z^2 - 1.531z + 0.5991)},$$

The closed-loop system is unstable, since the transfer function $T(z, \rho)$ has a pole at 1.024.

This issue has been discussed in Chap. 2, where it has been seen that to avoid this kind of trouble the reference model must possess the NMP zeros present in the process. Picking such a reference model would in principle require the identification of these zeros, in a partial identification phase prior to the controller's design. We would like to avoid model identification as one of the motivations for using a data-driven control design. Yet, only a partial identification is required, not the whole process model, and it can also be argued that identifying unstable singularities of a transfer function is less costly than identifying the stable ones [2, 14]. Thus this prior identification phase should be not such a big burden to the designer. In any case NMP zeros are a serious problem to begin with and the designer should not expect to get rid of it for free. Once the NMP zeros are known, they can be included in the model reference by the designer, and then a data-driven design with this reference model could be triggered.

But there is a more elegant alternative, that will be presented next. This alternative consists in embedding the identification of the zeros into the VRFT method itself, which is achieved through a simple modification of the cost function to be minimized. Prior identification is no longer needed and as a consequence there is less (if any) extra burden on the designer in order to successfully cope with possible NMP zeros of the plant.

3.4.1 The Flexible VRFT Criterion

If the reference model must contain the zeros of the plant, and these are unknown, the designer can not fix a priori the zeros of the reference model. Instead, a *flexible reference model* can be used

$$T(z,\eta) = \eta^T F(z), \tag{3.23}$$

where $\eta \in \mathbb{R}^q$ is a vector of free parameters and $F(z)$ is a q-vector of rational functions. By replacing the fixed reference model $T_d(z)$ by $T(z,\eta)$ and using the filter $L(z)$, the VRFT criterion (3.1) is changed into

$$J_0^{VR}(\eta,\rho) = \bar{E}\left\{L(z)\left[u(t) - \left(\frac{1-T(z,\eta)}{T(z,\eta)}C(z,\rho)\right)y(t)\right]\right\}^2. \tag{3.24}$$

In this formulation the denominator of the reference model is assigned, while the numerator is left free. If the number of free parameters q equals the order of the numerator of $T(z,\eta)$, then the numerator is entirely free and the formulation becomes conceptually equivalent to a pole assignment design. Then the optimization of the performance criterion $J_0^{VR}(\eta,\rho)$ with respect to η and ρ should "find" the zeros of the plant, and particularly the NMP zeros, along with the optimal controller parameters, all at once.

In the standard VRFT method, Assumption B_y is crucial. Our analysis for this new design criterion requires a similar assumption, which will state that there exists, within the class of reference models considered, one reference model for which controller matching is possible.

Assumption $\mathbf{B}_{\mathbf{NMP}}$ (Flexible matched control) There exists a pair (η_*, ρ_*) such that $J_0^{VR}(\eta_*, \rho_*) = 0$; equivalently

$$\exists \eta_*, \rho_* : \quad C(z, \rho_*) = \frac{T(z, \eta_*)}{[1 - T(z, \eta_*)]G(z)}. \tag{3.25}$$

Theorem 3.1 *Let Assumption* $\mathrm{B}_{\mathrm{NMP}}$ *be satisfied and all the poles of the controller be either inside or on the unit circle. Then the NMP zeros of* $G(z)$ *are also zeros of* $T(z, \eta_*)$.

Proof Let $G(z) = \frac{n_G(z)}{d_G(z)}$ and $T(z, \eta_*) = \frac{n_T(z,\eta_*)}{d_T(z)}$ be coprime factorizations of $G(z)$ and $T(z, \eta_*)$, where $n_G(z)$, $d_G(z)$, $n_T(z, \eta_*)$ and $d_T(z)$ are polynomials. From (2.9) we have

$$C(z, \rho_*) = \frac{n_T(z, \eta_*) d_G(z)}{[d_T(z) - n_T(z, \eta_*)] n_G(z)}. \tag{3.26}$$

By assumption, the denominator of $C(z, \rho_*)$ has no unstable roots (that is, roots outside the unit circle). Therefore, from (3.26), any unstable root of $n_G(z)$ must be canceled by a root of $n_T(z, \eta_*)$. □

According to Theorem 3.1, if $G(z)$ has NMP zeros and Assumption $\mathrm{B}_{\mathrm{NMP}}$ is satisfied, then the NMP zeros appear necessarily in the numerator of $T(z, \eta_*)$. Under these conditions, the global minimum of $J_0^{VR}(\eta, \rho)$ gives the desired result: a reference model containing the NMP zeros of the process and the controller that achieves this reference model. But this is a nasty cost function to minimize, as the decision variables appear even in its denominator, and VRFT is all about having a nice cost function to minimize so as to avoid iterative procedures.

In order to recover the nice numerical properties of the VRFT objective function, let us multiply $J_0^{VR}(\eta, \rho)$ by $T(z, \eta)$, arriving at another cost function:

$$\tilde{J}_0^{VR}(\eta, \rho) = \bar{E}[L(z)T(z, \eta)u(t) - L(z)C(z, \rho)(1 - T(z, \eta))y(t)]^2. \tag{3.27}$$

This new cost function must have the same minimum as the original one, which does happen under Assumption $\mathrm{B}_{\mathrm{NMP}}$:

$$\arg\min_{\eta,\rho} J_0^{VR}(\eta, \rho) = \arg \min_{\substack{\eta,\rho \\ (\eta,\rho) \neq \{0,0\}}} \tilde{J}_0^{VR}(\eta, \rho) = (\eta_*, \rho_*). \tag{3.28}$$

Notice however that because of the linear parametrization of both the controller and the reference model, $\tilde{J}_0^{VR}(0, 0) = 0$. Thus, the multiplication by $T(z, \eta)$ has created an additional—and undesired—global minimum at the origin of the parameter space. This is why the right hand side of (3.28) is subjected to a constraint that excludes this undesired minimum $(\eta, \rho) = \{0, 0\}$. This is a natural constraint, since $\eta = 0$ corresponds to the closed-loop behavior $T(z, \eta) = 0$, which does not make sense.

It now follows from Theorem 3.1 that, under Assumption B_{NMP}, the minimization (3.28) of $\tilde{J}_0^{VR}(\eta, \rho)$ yields a minimum (η_*, ρ_*) such that $T(z, \eta_*)$ contains all NMP zeros of $G(z)$. We have thus produced a data-based optimization problem whose solution detects the NMP zeros of the plant without utilizing a full order model identification procedure, the only assumption being that the controller structure is such that the desired closed-loop poles can be achieved. But the resulting cost function is no longer quadratic in the parameter, so it remains to show how to best perform its optimization.

3.4.2 Implementation Issues

Inserting (3.23) and $C(z, \rho) = \rho^T \bar{C}(z)$ into (3.27) yields

$$\tilde{J}_0^{VR}(\eta, \rho) = \bar{E}\{\eta^T F(z)[u_L(t) + \rho^T \bar{C}(z) y_L(t)] - \rho^T \bar{C}(z) y_L(t)\}^2. \quad (3.29)$$

The argument in (3.29) is bilinear in η and ρ. For objective functions with this structure, the minimization can be treated as a sequence of least squares problems, as described below [12]. For a fixed η_1 the minimization $\tilde{J}_0^{VR}(\eta_1, \rho)$ in ρ can be solved by least squares, obtaining an optimal value—say ρ_1; then, with the value of ρ_1 just obtained, optimize the resulting cost function $\tilde{J}_0^{VR}(\eta, \rho_1)$, obtaining a new value η_2. Proceeding iteratively in this way, at each iteration the following pair of least squares problem must be solved

$$\eta_i = \arg\min_{\eta} \tilde{J}_0^{VR}(\eta, \rho_{i-1}), \quad (3.30)$$

$$\rho_i = \arg\min_{\rho} \tilde{J}_0^{VR}(\eta_i, \rho), \quad (3.31)$$

where i is the iteration number. Each one of these least squares problems at iteration i has the explicit solutions:

$$\begin{aligned}
\eta_i &= \bar{E}\{[F(z)w(\rho_{i-1}, t)][F(z)w(\rho_{i-1}, t)]^T\}^{-1} \\
&\quad \times \bar{E}\{[F(z)w(\rho_{i-1}, t)][C(z, \rho_{i-1})L(z)y(t)]\}, \quad (3.32) \\
\rho_i &= \bar{E}\{[\bar{C}(z)v(\eta_i, t)][\bar{C}(z)v(\eta_i, t)]^T\}^{-1} \\
&\quad \times \bar{E}\{[\bar{C}(z)v(\eta_i, t)][M(z, \eta_i)L(z)u(t)]\}, \quad (3.33) \\
w(\rho, t) &\triangleq L(z)[u(t) + \rho^T \bar{C}(z)y(t)], \\
v(\eta, t) &\triangleq L(z)[1 - \eta^T F(z)]y(t).
\end{aligned}$$

This sequential least squares algorithm is guaranteed to converge at least to a local minimum [12, 17]. An initialization must be provided for $T(z, \eta)$; one possible choice is to use $T(z, \eta_0) = T_d(z)$. This is an iterative algorithm, but it is very important to notice right away that this algorithm is NOT iterative in the sense (most commonly used in data-driven design) that iterative adjustments of the controller's parameters are made. The controller's parameters are set only once, at the end of the sequential least squares algorithm, and the data from the system are collected only once. The "one-shot" property of the VRFT method is kept intact.

3.4.2.1 Remarks on the Filter $L(z)$

A few remarks are in order concerning the filter $L(z)$. First, one may ask why the filter is used at all, since we consider that matching Assumption B_{NMP} is satisfied, and the filter is necessary only when the matching condition is violated. But at each iteration a "classical" VRFT is performed with the current reference model $T(z, \eta_i)$, for which the matching condition B_y is not satisfied. Assumption B_{NMP} only guarantees that Assumption B_y is satisfied for $\eta = \eta_*$. This is the reason why a filter $L(z)$ must be used at each iteration.

This filter's job is to approximate the minima of the reference tracking $J_y(\rho)$ and the VRFT criterion $J^{VR}(\rho)$ and to do that it must be calculated by an expression involving the reference model. In the iterative algorithm involving the flexible criterion, at each iteration the reference model $T(z, \eta)$ is different, and thus so must be the filter, which now must be calculated from

$$|L(e^{J\omega})|^2 = |1 - T(e^{J\omega}, \eta)|^2 |T(e^{J\omega}, \eta)|^2 \frac{\Phi_r(\omega)}{\Phi_u(\omega)}, \quad \forall\omega \in [-\pi; \pi]. \quad (3.34)$$

3.4.3 Two-Step Procedure

The global minimum of the flexible criterion $\tilde{J}_0^{VR}(\eta, \rho)$ corresponds to a reference model that contains the NMP zeros of the plant, if any. But we don't know a priori whether there are NMP zeros or not, and when there are not it is preferable to use the standard VRFT method, in which the designer choses the reference model at will. With that in mind, and in order to keep the designer's options as broad as possible, the following two step procedure can be applied to processes where NMP zeros may or may not exist.

Let $T_d(z)$ be the designer's choice for the reference model, that he/she will not use directly in a VRFT design because there may be NMP zeros in the process. Then proceed in the following two steps.

Step 1 Pick a flexible reference model $T(z, \eta)$ with the same poles as $T_d(z)$ and minimize $\tilde{J}_0^{VR}(\eta, \rho)$. Let $(\hat{\eta}, \hat{\rho})$ be the minimizing parameters just found and verify whether or not the performance provided by the resulting closed-loop—given by $T(z, \hat{\eta})$—is satisfactory. If it is, apply $C(z, \hat{\rho})$ to the system; if not, go to Step 2.

Step 2 If $T(z, \hat{\eta})$, obtained in Step 1, has NMP zeros, then modify the reference model $T_d(z)$ so that it contains these NMP zeros, keeping its remaining singularities, and solve a standard VRFT design with this new reference model. If there are not

NMP zeros, just solve a standard VRFT design with the initially chosen reference model $T_d(z)$.

In case Step 1 has determined that there are indeed NMP zeros, these must be included in the reference model. The proposed modification of the original reference model $T_d(z)$ to include these zeros is a way to specify a performance that is as close as possible to the originally desired performance $T_d(z)$, given the constraints imposed by the NMP zeros.

3.5 The Noisy Case

The VRFT formulations just presented were derived for noiseless processes, an ideal condition that is rarely found. In this section, we will analyze what happens when the VRFT method is applied to noisy systems and present a modification to it that broadens its application to this case.

The optimal parameter $\hat{\rho}$ provided by the VRFT method is calculated by means of (3.10). In the presence of measurement noise, this is no longer a deterministic quantity, but rather a random variable, which represents a stochastic estimate of the real parameter value ρ_d. The relevant properties of the VRFT method are the statistical properties of the random variable $\hat{\rho}$. These properties are studied in the sequel.

The output of the process is given by

$$y(t) = G(z)u(t) + H(z)e(t). \tag{3.35}$$

An inverse model can be formed isolating $u(t)$ in (3.35):

$$\mathcal{S}_{inv} : \; u(t) = \frac{1}{G(z)} y(t) - \frac{1}{G(z)} H(z)e(t). \tag{3.36}$$

But from the definition of the ideal controller $C_d(z)$ in (2.9), the process' transfer function $G(z)$ can be written as a function of $C_d(z)$ and $T_d(z)$:

$$\frac{1}{G(z)} = C_d(z)\frac{1 - T_d(z)}{T_d(z)}. \tag{3.37}$$

Substituting (3.37) in (3.36) gives

$$\mathcal{S}_{inv} : \; u(t) = C_d(z)\frac{1 - T_d(z)}{T_d(z)} y(t) - \frac{1}{G(z)} H(z)e(t), \tag{3.38}$$

$$u(t) = C_d(z)\bar{e}(t) - \frac{1}{G(z)} H(z)e(t). \tag{3.39}$$

The VRFT method can be seen as a prediction error approach to the identification of this system using the input-output data $\bar{e}(t)$ and $u(t)$, a fact that has already been mentioned earlier, in this chapter's Introduction. The VRFT method identifies for this system a model of the form

$$\hat{u}(t) = \rho^T \varphi(t) \tag{3.40}$$

where the regressor vector $\varphi(t)$ is defined as

$$\varphi(t) = \bar{C}(z)\bar{e}(t).$$

Posed like this, the control design via VRFT is a standard identification problem with a linear regressor, and it will be treated as such. The random variable $\hat{\rho}$ presents an error regarding the real optimal parameter value ρ_d which is conveniently decomposed into two terms:

$$\hat{\rho} - \rho_d = (\hat{\rho} - \rho_*) + (\rho_* - \rho_d) \tag{3.41}$$

where ρ_* is the asymptotic value of the estimate $\hat{\rho}$, that is,

$$\rho_* = \lim_{N\to\infty} \left[\sum_{t=1}^{N} \varphi(t)\varphi^T(t)\right]^{-1} \sum_{t=1}^{N} \varphi(t)u(t) = \bar{E}[\varphi(t)\varphi^T(t)]^{-1}\bar{E}[\varphi(t)u(t)]$$

as defined previously in (3.9).

The first term $\hat{\rho} - \rho_*$ in (3.41) is a random quantity called the *variance error*; its origin is the finiteness of the data set used to obtain the estimate. The second term $\rho_* - \rho_d$ is a deterministic quantity called *bias error*, and is inherent to the problem formulation. An estimate is said to be consistent if the bias term is zero. A consistent estimate can be made arbitrarily close to the real optimal parameter value ρ_d by collecting more information—that is, more data—because the variance error tends to zero as N tends to infinity. An estimate that is not consistent can not be arbitrarily improved; collecting more data will not change the second term.

It is therefore a minimal requirement in most estimation problems in statistics, and parameter identification is not different, that the solution must be consistent. The following theorem characterizes the consistency of the estimate $\hat{\rho}$.

Theorem 3.2 *Let Assumption* B_y *be satisfied. The estimate* (3.9) *of the ideal controller parameters* ρ_d *produced by the VRFT method is* biased *by the following quantity.*

$$bias = \bar{E}[(\varphi(t)\varphi^T(t))^{-1}\varphi(t)G^{-1}(z)H(z)e(t)]. \tag{3.42}$$

Proof The controller's output—the signal $u(t)$—is given by (3.39). Substituting (3.39) into (3.9) yields

$$\begin{aligned}
\rho_* &= \bar{E}[\varphi(t)\varphi^T(t)]^{-1}\bar{E}\left[\varphi(t)\left(\varphi^T(t)\rho_d - \frac{1}{G(z)}H(z)e(t)\right)\right] \\
&= \bar{E}[\varphi(t)\varphi^T(t)]^{-1}\bar{E}[\varphi(t)\varphi^T(t)]\rho_d - \bar{E}[\varphi(t)\varphi^T(t)]^{-1}\bar{E}[\varphi(t)G^{-1}(z)H(z)e(t)] \\
&= \rho_d - \bar{E}[\varphi(t)\varphi^T(t)]^{-1}\bar{E}[\varphi(t)G^{-1}(z)H(z)e(t)].
\end{aligned}$$

□

A consistent estimate of the ideal parameter ρ_d is obtained if the bias term is zero, which will happen if and only if

$$\bar{E}[\varphi(t)G^{-1}(z)H(z)e(t)] = 0. \tag{3.43}$$

Since the regressor vector is formed by past values of $y(t)$, which are correlated to past values of the noise, it is not easy to imagine meaningful conditions under which (3.43) would be satisfied. So, the standard VRFT provides in almost all real situations a biased estimate, which is a serious drawback, unless the noise level is negligible. To overcome this limitation, an instrumental variable (IV) approach to the identification of $C_d(z)$ can be used.

An instrumental variable is a p-vector $\zeta(t)$ satisfying

$$\bar{E}[\zeta(t)\varphi(t)^T] > 0$$

and

$$\bar{E}[\zeta(t)e(s)^T] = 0 \quad \forall t, s.$$

Once an appropriate IV has been found, the controller parameters are given by:

$$\rho^{IV} = \bar{E}[\zeta(t)\varphi(t)^T]^{-1}\bar{E}[\zeta(t)u(t)]. \tag{3.44}$$

It is also standard knowledge in identification theory that the estimate obtained with this IV approach is unbiased, as can be shown through the same argument used in the proof of Theorem 3.2. This IV approach actually is the form of VRFT that is standard practice, because it removes the bias and maintains the other properties of the VRFT method. As for the determination of the IV itself, different choices exist, as discussed next.

3.5.1 Choosing the Instrumental Variable

A traditional way to generate instrumental variables in identification theory is to substitute the output $y(t)$ in the regressor vector by a noise-free approximation. Applying this idea to the regressor vector of VRFT, which is given by

$$\varphi(t) = \bar{C}(z)\left(\frac{1 - T_d(z)}{T_d(z)}\right)y(t),$$

results in the following family of IVs:

$$\zeta(t) = \bar{C}(z)\left(\frac{1 - T_d(z)}{T_d(z)}\right)y'(t), \tag{3.45}$$

where $y'(t)$ is the "noise-independent output", that is, a signal that is generated similarly to the output $y(t)$ but without the influence of the noise measured in the experiment. This will make the IV uncorrelated to the noise, but strongly correlated to the regressor vector, as desired. Equation (3.45) defines a whole family of IVs rather than a single one, because the "noise-independent output" $y'(t)$ may be generated in a number of different ways.

Two different ways to obtain $y'(t)$ have been proposed for VRFT when the data are collected in an open-loop experiment [5]. The first one consists of repeating the same experiment twice and guarantees that asymptotically $\rho^{IV} = \rho_*$. However, an additional experiment on the plant is required. The second one consists in estimating a rough model for the process with the data collected and then simulating this model to generate the IV. This procedure does not guarantee that $\rho^{IV} = \rho_*$, but the bias tends to be small and it does not require an additional experiment. The procedures for obtaining these two different IVs are detailed below.

- *Repeated experiment.* Perform a second experiment on the plant using the same input $\{u(t)\}_{t=1,\dots,N}$ and collect the corresponding output sequence $\{y'(t)\}_{t=1,\dots,N}$. Then construct the instrumental variable through (3.45). Notice that $\{y'(t)\}_{t=1,\dots,N}$ is different from $\{y(t)\}_{t=1,\dots,N}$, since the two sequences are affected by two different realizations of the noise in the two experiments. If we assume, as it is reasonable, that the noise signals in the two experiments are uncorrelated, then ρ^{IV} is an unbiased estimate of ρ_d.
- *Identification of the plant.* Identify a model $\hat{G}(z)$ of the plant from the set of data $\{u(t), y(t)\}_{t=1,\dots,N}$ and generate the simulated output $y'(t) = \hat{G}(z)u(t)$. Then construct the instrumental variable through (3.45). Due to the inaccuracy of the estimated $\hat{G}(z)$, which also depends on the noise, this second method does not guarantee that the estimate ρ^{IV} asymptotically tends to ρ_d. It does reduce the bias, but not necessarily to zero.

3.5.1.1 Instrumental Variables for the Flexible VRFT Criterion

In the presence of noise, the optimization of the flexible VRFT criterion through the sequential Least Squares (3.32) and (3.33) will of course suffer from the same bias problems described for the standard VRFT method. On the other hand, the solution provided by the instrumental variable approach can be applied also to the flexible VRFT criterion, *mutatis mutandis*. In the sequel we detail the use of the instrumental variable obtained by means of the repeated experiment, similarly to the one presented above for the optimization of the standard VRFT criterion.

Concerning the case of an open-loop experiment, in the first experiment we collect the input data $u(t)$ and the output data $y(t)$; in the second one, only the output data $y'(t)$ must be collected, because the input data $u(t)$ is the same as in the first experiment. The regression vector used to calculate η^i in the flexible VRFT criterion is as given in (3.32):

$$\begin{aligned}\varphi_\eta(t,\rho) &= F(z)w(t,\rho)\\ &= F(z)L(z)[u(t)+\rho^T\bar{C}(z)y(t)].\end{aligned} \tag{3.46}$$

Then the instrumental variable is constructed by simple substitution of the output $y(t)$ by a noise-independent parallel $y'(t)$:

$$\begin{aligned}\xi_\eta(t,\rho) &= F(z)w^{IV}(t,\rho)\\ &= F(z)L(z)[u(t)+\rho^T\bar{C}(z)y'(t)].\end{aligned}\tag{3.47}$$

If a closed-loop experiment is used instead, then $u(t)$ is also correlated with the noise. Then to obtain an IV that is not correlated to the noise, the input $u(t)$ must also be substituted by a noise-independent signal $u'(t)$, which will be the process' input in the second experiment. The instrumental variable for closed-loop experiments must be computed as

$$\xi_\eta(t,\rho) = F(z)L(z)[u'(t)+\rho^T\bar{C}(z)y'(t)].$$

Analogously, in order to compute ρ_i, the regression vector is given by

$$\begin{aligned}\varphi_\rho(t,\eta) &= \bar{C}(z)v(t,\eta)\\ &= \bar{C}(z)L(z)[1-\eta^T F(z)]y(t).\end{aligned}\tag{3.48}$$

Hence, the instrumental variable for this sub-problem is given by

$$\begin{aligned}\xi_\rho(t,\eta) &= \bar{C}(z)v^{IV}(t,\eta)\\ &= \bar{C}(z)L(z)[1-\eta^T F(z)]y'(t),\end{aligned}\tag{3.49}$$

where $y'(t)$ is obtained as in the previous case. In this case, it is not relevant whether the experiment is performed in open or closed loop, since $v(t,\eta)$ does not depend on $u(t)$.

Thus, in the noisy case, each step of the algorithm (3.30)–(3.31) is solved through the following equations:

$$\begin{aligned}\eta_i^{IV} &= \bar{E}\{[\xi_\eta(t,\rho_{i-1})\varphi_\eta^T(t,\rho_{i-1})]^{-1}\xi_\eta(t,\rho_{i-1})[C(z,\rho_{i-1})L(z)y(t)]\}\\ &= \bar{E}\{[F(z)w^{IV}(\rho_{i-1},t)][F(z)w(\rho_{i-1},t)]^T\}^{-1}\\ &\quad\times \bar{E}\{[F(z)w^{IV}(\rho_{i-1},t)][C(z,\rho_{i-1})L(z)y(t)]\},\end{aligned}\tag{3.50}$$

$$\begin{aligned}\rho_i^{IV} &= \bar{E}\{[\xi_\rho(t,\eta_i)\varphi_\rho^T(t,\eta_i)]^{-1}\xi_\rho(t,\eta_i)[T(z,\eta_i)L(z)u(t)]\}\\ &= \bar{E}\{[\bar{C}(z)v^{IV}(\eta_i,t)][\bar{C}(z)v(\eta_i,t)]^T\}^{-1}\\ &\quad\times \bar{E}\{[\bar{C}(z)v^{IV}(\eta_i,t)][T(z,\eta_i)L(z)u(t)]\}.\end{aligned}\tag{3.51}$$

3.6 Case Studies

A few simulated case studies will be presented in order to illustrate the main issues concerning the application of the VRFT method in different situations. In the first example, we explore the need of instrumental variables when the signals are corrupted by noise, illustrating the variance and bias properties of the estimates. In the second, we show how to use the flexible criterion in order to estimate the NMP zeros of the process, if there are any, and then how to use this information when designing the controller. Experimental studies further explaining and illustrating the application of VRFT will be presented in Chap. 8.

3.6.1 The Need of Instrumental Variables

The following example will illustrate that the use of instrumental variables is necessary to provide a meaningful controller, except for very low noise levels. Suppose that the process to be controlled is described by

$$G(z) = \frac{1}{z - 0.9}, \qquad H(z) = 1. \tag{3.52}$$

Once again, the following reference model of order one will be used:

$$T_d(z) = \frac{1-a}{z-a} \tag{3.53}$$

where a is the pole of the desired closed-loop transfer function. The controller is a PI:

$$C(z,\rho) = \frac{\varrho_1 z + \varrho_2}{z-1}.$$

This linear parameterization is used in all computations, but to facilitate an intuitive visualization and interpretation, the results are presented in terms of the controllers gain and zero instead, that is, in terms of ϖ_1 and ϖ_2 in the following equation.

$$C(z,\rho) = \frac{\varpi_1 (z + \varpi_2)}{z-1}. \tag{3.54}$$

The ideal controller is easily calculated, and it belongs to the controller class for any value of a:

$$C_d(z) = \frac{(1-a)(z-0.9)}{z-1}. \tag{3.55}$$

Since the output signal is corrupted by noise, the VRFT requires the use of instrumental variables to produce unbiased estimates. The basic VRFT formulation, that is, the one with standard least squares estimate, yields a biased result. This example will illustrate these consistency properties, also from a quantitative point of view.

Results for different signal to noise ratios and different reference models, given by (3.53) with different values of the parameter a, will be presented. For each case

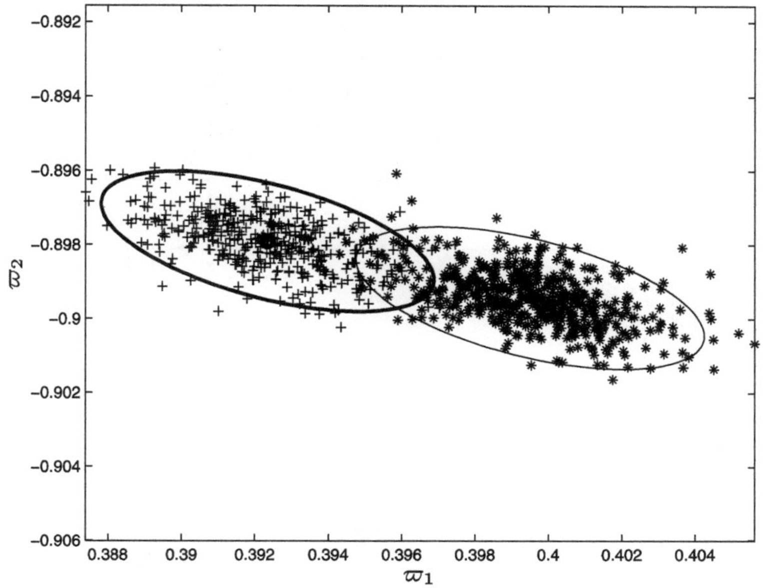

Fig. 3.4 Estimated parameters obtained through LS (+) and IV (∗) from 500 Monte Carlo runs where $\sigma_e^2 = 0.01$. Ellipse plots of 95% confidence interval. The ideal parameter vector is $\rho_d = [0.4 \;\; -0.9]^T$

we ran 500 Monte Carlo simulations, collecting the open-loop input signal $u(t)$, which is a PRBS[4] with amplitude ± 1, and the corresponding output signal $y(t)$, which is corrupted by white noise. Each Monte Carlo run consisted of $N = 1,000$ data samples.

Start with a noise variance $\sigma_e^2 = 0.01$, and a reference model with $a = 0.6$. The Monte Carlo experiments were ran and for each run the controller parameters were calculated using VRFT with both the standard least squares method—LS—and the instrumental variable (two experiments with the same input)—IV. Figure 3.4 shows the estimated parameter (the controller gain ϖ_1 and the controller zero ϖ_2) obtained in each run, for both methods: LS, marked as + and IV, marked as ∗. We also plotted the resulting 95% confidence ellipse of each method, which shows the covariance of the estimate obtained in each case.[5]

Each ellipse is centered at its mean value, and these mean values are given in Table 3.5. Substituting $a = 0.6$ into (3.55), the ideal value of the parameter vector is

$$\rho_d = \begin{bmatrix} 0.4 \\ -0.9 \end{bmatrix}.$$

[4]Pseudo Random Binary Signal.

[5]The ellipses were computed from the data, so this is the sample covariance.

Table 3.5 Mean values of $\hat{\rho}$ for LS and IV when $a = 0.6$ and $\sigma_e^2 = 0.01$

Method	$\hat{\rho}_m^T$	$\|(\hat{\rho}_m - \rho_d)\|$
Ideal	[0.4000 −0.9000]	–
LS	[0.3922 −0.8978]	0.008125
IV	[0.3993 −0.8993]	0.000988

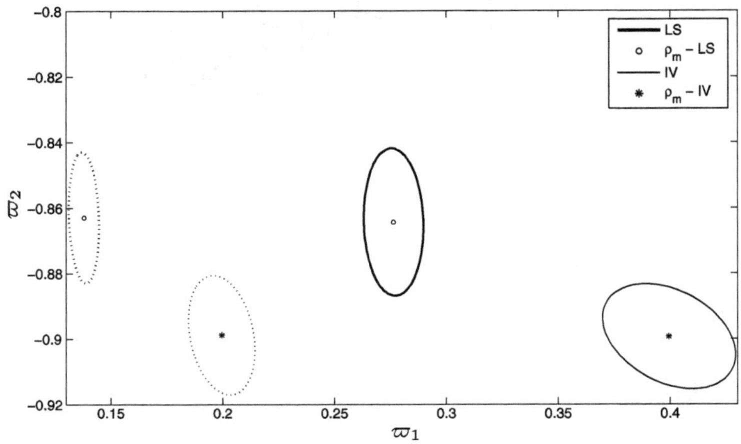

Fig. 3.5 95% confidence ellipses obtained from the estimates using LS (*thicker lines*, centered at ∘) and IV (*thinner lines*, centered at ∗) for $a = 0.6$ (*solid lines*) and $a = 0.8$ (*dotted lines*). The LS is strongly biased

The controller parameters are closer to the "correct" value ρ_d when calculated with the use of IV, but the standard least squares solution also gives very close approximations, because the noise level is low. Although the standard least squares solution provides a biased estimate, this bias is small and could be acceptable in a practical application. On the other hand, it is also observed in Fig. 3.4 that the estimate variances obtained with both approaches have about the same magnitude.

Consider now a higher level of noise: $\sigma_e^2 = 0.25$. For this noise level we want to show the statistical properties of the estimates—through LS and IV—for different reference models, that is, for different values of a. Figure 3.5 shows the 95% confidence ellipses obtained from the estimates using both methods for two different values of a: 0.6 and 0.8. The ellipses plotted with solid lines are the ones obtained with $a = 0.6$, while the ones plotted with dotted lines correspond to $a = 0.8$. The thinner line ellipses with the central point denoted with an asterisk are the ones obtained from IV estimates, while the thicker ellipses centered at a circle are the ones obtained with LS. The mean values of each estimate ($\hat{\rho}_m$) are presented in Table 3.6, along with their Euclidean distances to the ideal controller parameter. A negligible distance between the mean value $\hat{\rho}_m$ and the ideal controller parameter ρ_d indicates that the estimate is unbiased.

Table 3.6 Mean values of the parameters $\hat{\rho}$ obtained for $a = 0.6$, $a = 0.8$ and $\sigma_e^2 = 0.25$

a	Method	$\hat{\rho}_m^T$	$\|(\hat{\rho}_m - \rho_d)\|$
	Ideal	[0.2000 −0.9000]	–
0.8	LS	[0.1379 −0.8629]	0.072302
	IV	[0.1995 −0.8988]	0.001277
	Ideal	[0.4000 −0.9000]	–
0.6	LS	[0.2764 −0.8643]	0.128634
	IV	[0.4006 −0.8991]	0.001043

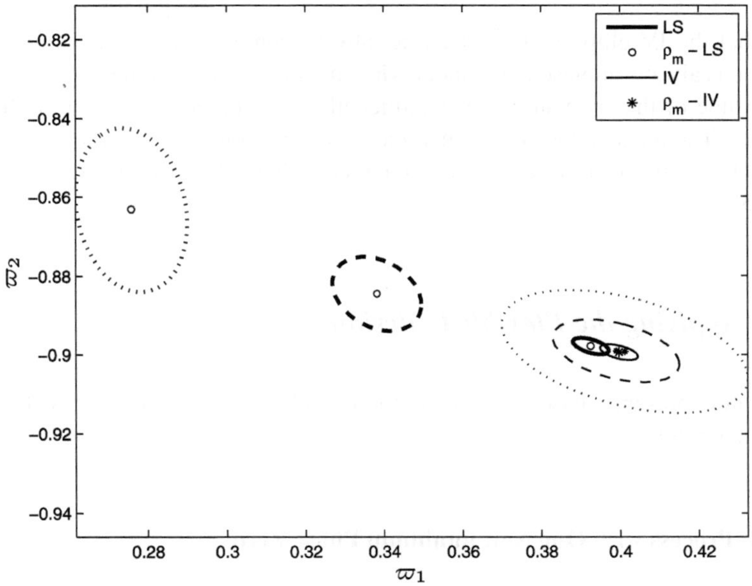

Fig. 3.6 95% confidence ellipses, obtained with LS (*thicker lines* and centered at ∘) and IV (*thinner lines* centered at ∗) for $\sigma_e^2 = 0.01$ (*solid lines*), $\sigma_e^2 = 0.10$ (*dashed lines*) and $\sigma_e^2 = 0.25$ (*dotted lines*)

In both designs ($a = 0.6$ and $a = 0.8$), the estimates obtained with IV are unbiased, while the ones obtained with LS are biased. Note that the bias of the LS estimate is quite significant in both cases, being larger with the faster reference model $a = 0.6$. The variances of both the IV and LS estimates are also larger for the faster reference model.

Finally, consider yet a third scenario, where the reference model pole is fixed at $a = 0.6$ and noise level varies. The 95% confidence ellipses for three different values of σ_e^2 are plotted in Fig. 3.6. Again, ellipses plotted with thicker lines and centered at a circle are obtained with LS while the thinner ones centered at an asterisk are obtained with IV. The solid line ellipses were obtained for $\sigma_e^2 = 0.01$, dashed line ellipses are for $\sigma_e^2 = 0.10$ and dotted lines ellipses are obtained for $\sigma_e^2 = 0.25$.

Table 3.7 Mean values of the parameters $\hat{\rho}$ obtained for $a = 0.6$, $\sigma_e^2 = 0.01$, $\sigma_e^2 = 0.10$ and $\sigma_e^2 = 0.25$

σ_e^2	Method	$\hat{\rho}_m^T$	$\|(\hat{\rho}_m - \rho_d)\|$
	Ideal	[0.4000 −0.9000]	–
0.01	LS	[0.3922 −0.8978]	0.008125
0.01	IV	[0.3993 −0.8993]	0.000988
0.10	LS	[0.3383 −0.8847]	0.063547
0.10	IV	[0.3993 −0.8993]	0.000986
0.25	LS	[0.2764 −0.8643]	0.128633
0.25	IV	[0.4006 −0.8991]	0.001043

Note that the IV ellipses are concentric, always centered at ρ_d, meaning that the estimate is always unbiased, no matter what the noise level is. When we use LS, the mean value of the estimate varies significantly with the noise level, as well as its variance. The mean values for each value of σ_e^2 tested for both methods as well as the Euclidean distances of each one to the ideal value of the parameter are shown in Table 3.7.

3.6.2 Applying the Flexible Criterion

Let us now present a case study using the flexible VRFT scheme for NMP plants presented in Sect. 3.4.

3.6.2.1 Process with One Non-minimum Phase Zero

Let the transfer function of the process be

$$G(z) = \frac{(z-1.2)(z-0.4)}{z(z-0.3)(z-0.8)}. \tag{3.56}$$

We want to control this process with a PID controller

$$C(z,\rho) = \rho^T \bar{C}(z) = [\varrho_1\ \varrho_2\ \varrho_3] \begin{bmatrix} \frac{z^2}{z^2-z} \\ \frac{z}{z^2-z} \\ \frac{1}{z^2-z} \end{bmatrix}. \tag{3.57}$$

The data are collected from a closed-loop experiment, where a unit step is applied as the reference signal and the controller in the loop is given by

$$C_0(z) = \frac{-0.7(z-0.4)(z-0.6)}{z^2-z}. \tag{3.58}$$

Assumption B_{NMP} Is Satisfied Consider the following reference model, which has been chosen in the absence of any knowledge on the NMP zero of $G(z)$ and for which Assumption B_y is not satisfied:

$$T_d(z) = \frac{0.0706z^2}{(z-0.885)(z^2-0.706z+0.32)}. \tag{3.59}$$

The standard VRFT criterion used with this reference model yields the controller

$$C(z,\hat{\rho}) = \frac{-2.269(z^2-1.655z+0.7007)}{z^2-z},$$

which causes the closed loop to be unstable, as can be seen in the resulting closed-loop transfer function

$$T(z,\hat{\rho}) = \frac{-2.2693(z-1.200)(z-0.4000)(z^2-1.655z+0.7007)}{(z-0.3909)(z^2-1.774z+0.7905)(z^2-2.204z+2.470)}.$$

The system instability results from the fact that the NMP zero present in $G(z)$ is not in the reference model $T_d(z)$; as a consequence, the ideal controller would provide an internally unstable closed loop by canceling the NMP zero, which causes closed-loop instability even though the ideal controller can not be exactly achieved (Assumption B_y is not satisfied). We thus use the two-step procedure with the following flexible reference model, which satisfies Assumption $\mathrm{B}_{\mathrm{NMP}}$ and has the same poles as the desired reference model:

$$T(z,\eta) = \frac{\vartheta_1 z^2+\vartheta_2 z+\vartheta_3}{(z-0.885)(z^2-0.706z+0.32)}. \tag{3.60}$$

We minimize $\tilde{J}_0^{VR}(\eta,\rho)$ with respect to η and ρ using the iterative procedure (3.30)–(3.31). The step responses of the fixed reference model $T(z,\hat{\eta}_i)$ and the closed loop $T(z,\hat{\rho}_i)$ obtained at iterations 1 and 30 are presented in Fig. 3.7. Note that $T(z,\hat{\eta}_{30})$ and $T(z,\hat{\rho}_{30})$ are almost indistinguishable. Table 3.8 shows the evolution of the corresponding parameters, by means of the numerators of the controller and the flexible reference model, obtained in different iterations. The transfer functions $T(z,\hat{\eta}_{30})$ and $C(z,\hat{\rho}_{30})$ at iteration 30 are as follows:

$$T(z,\hat{\eta}_{30}) = \frac{-0.5908(z-1.200)(z-0.4022)}{(z-0.885)(z^2-0.706z+0.32)},$$

$$C(z,\hat{\rho}_{30}) = \frac{-0.5903(z-0.8000)(z-0.3004)}{z^2-z}.$$

Observe in Table 3.8 that $T(z,\hat{\eta}_{30})$ reproduces both zeros of $G_1(z)$ with a good precision, and the controller $C(z,\hat{\rho}_{30})$ is such that its zeros cancel the poles of the process. Note also that a good estimate of the NMP zero is already present at iteration $i=20$, while convergence to the minimum phase zero is slower. This observation is consistent with the findings in the literature, which show that NMP zeros are

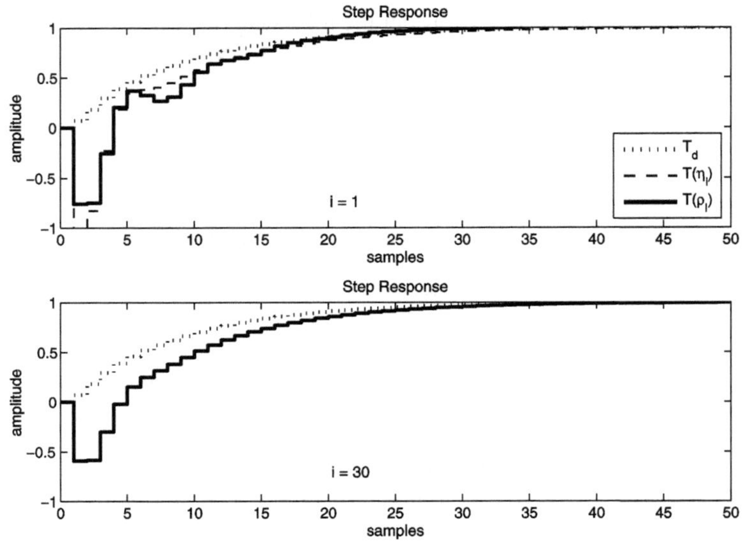

Fig. 3.7 Step responses of the fixed reference model (3.59) $T_d(z)$, the flexible model $T(z, \hat{\eta}_i)$ and the closed-loop system $T(z, \hat{\rho}_i)$ for process $G_1(z)$ for iterations 1 and 30

Table 3.8 Evolution of $T(z, \eta)$, $C(z, \rho)$ (by means of their numerators—gains and zeros) and $\tilde{J}_0^{VR}(\hat{\eta}, \hat{\rho})$ in the iterative procedure for the process $G_1(z)$

i	num($T(z, \hat{\eta}_i)$)	$\tilde{J}_0^{VR}(\hat{\eta}_i, \hat{\rho}_{i-1})$ ($\times 10^{-6}$)
1	$-1.078(z-1.182)(z-0.6404)$	1.6772616
2	$-0.8782(z-1.186)(z-0.5686)$	6.4634327
10	$-0.6307(z-1.196)(z-0.4290)$	0.0104166
20	$-0.5962(z-1.199)(z-0.4061)$	0.0002024
30	$-0.5908(z-1.200)(z-0.4022)$	0.0000064

i	num($C(z, \hat{\rho}_i)$)	$\tilde{J}_0^{VR}(\hat{\eta}_i, \hat{\rho}_i)$ ($\times 10^{-6}$)
1	$-0.7602(z-0.7448)(z-0.5292)$	12.928054
2	$-0.7117(z-0.7704)(z-0.4509)$	1.8210199
10	$-0.6109(z-0.7963)(z-0.3245)$	0.0086147
20	$-0.5931(z-0.7996)(z-0.3038)$	0.0001776
30	$-0.5903(z-0.8000)(z-0.3004)$	0.0000057

easier to estimate than minimum phase zeros [13, 14]. This design is by itself satisfactory and shows the efficiency of the flexible design criterion in coping with NMP zeros. Whereas a standard VRFT design would lead to an unstable closed loop, with the proposed design approach the closed loop is stable and its behavior resembles the desired one, specified by the fixed reference model.

We can however make the closed-loop behavior even closer to that of the fixed reference model. Indeed, in applying the flexible reference model we have left both

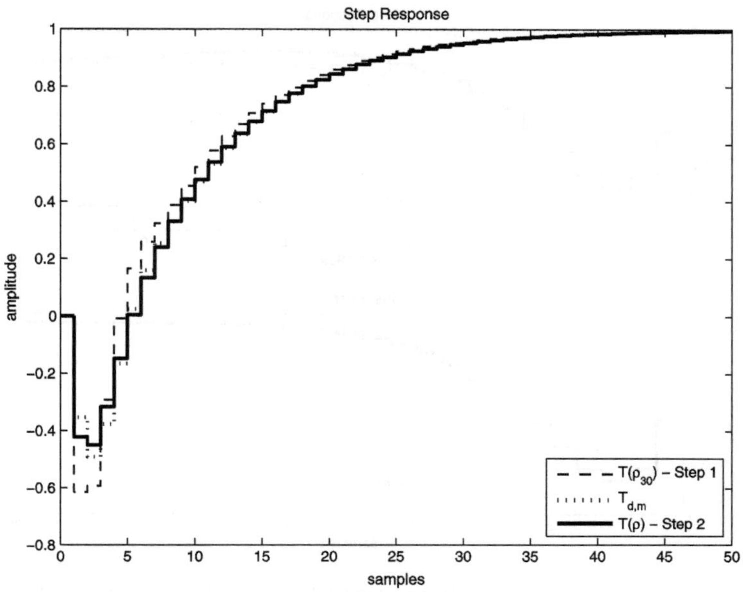

Fig. 3.8 Closed-loop responses $T(z,\hat{\rho})$ obtained at the end of Steps 1 and 2, compared to the modified reference model (3.61)

zeros of the process unchanged in the closed-loop transfer function. But only one of these zeros is NMP and thus needs to be there; the other closed-loop zero can still be assigned by the designer. So, once we know that the process actually possesses a NMP zero and its location in the complex plane, we change $T_d(z)$ to include this NMP zero and then use the standard VRFT. This new reference model is defined by

$$T_{d_m}(z) = \frac{-0.3532(z-1.200)z}{(z-0.885)(z^2-0.706z+0.32)}, \tag{3.61}$$

where the gain is chosen so that $T_{d_m}(1) = 1$. The standard VRFT method using (3.61) yields the controller

$$C(z,\hat{\rho}) = \frac{-0.4516(z-0.8033)(z-0.097992)}{z^2-z}.$$

Figure 3.8 shows the step responses obtained at the end of Step 1 with $T(z,\hat{\rho}_{30}) = T(z,\hat{\eta}_{30})$ and at the end of Step 2 with $T(z,\hat{\rho})$ where $\hat{\rho}$ is calculated from (3.10), as well as the step response of $T_{d_m}(z)$. The responses of $T(z,\hat{\rho})$ and $T_{d_m}(z)$ are very similar, but in Step 2 a smaller inverse response is obtained.

Assumption B_{NMP} Is Not Satisfied In the previous example, the reference model (3.60) was chosen in such a way that the matching condition (3.25) is satisfied for some (η_*, ρ_*) pair. Since the process $G(z)$ is unknown, it can not be guaranteed that the designer can choose the poles of $T(z,\eta)$ such that Assumption B_{NMP} is satisfied. Let us see how the design method behaves in this situation.

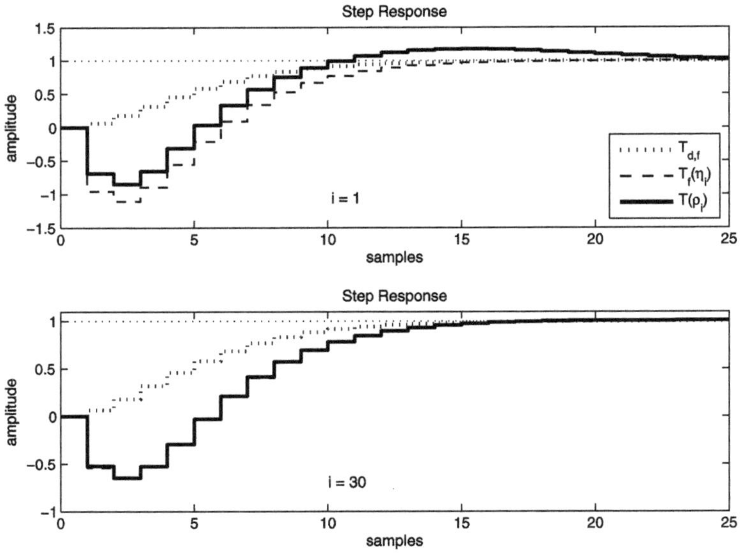

Fig. 3.9 Step responses of the fixed reference model $T_{d_f}(z)$ (3.62), the flexible model $T_f(z, \hat{\eta}_i)$ and the closed-loop system $T(z, \hat{\rho}_i)$ for process $G_1(z)$ for iterations 1 and 30 in Step 1

Suppose now that we choose for the same process (3.56) a different fixed reference model

$$T_{d_f}(z) = \frac{0.064z^2}{(z-0.6)^3}, \tag{3.62}$$

where the subscript f denotes "faster", as well as a flexible one defined as

$$T_f(z, \eta) = \frac{\vartheta_1 z^2 + \vartheta_2 z + \vartheta_3}{(z-0.6)^3},$$

for which Assumption B_{NMP} is not satisfied. In Step 1 we obtain, in 30 iterations,

$$T_f(z, \hat{\eta}_{30}) = \frac{-0.5422(z-1.197)(z-0.4021)}{(z-0.6)^3},$$

$$C(z, \hat{\rho}_{30}) = \frac{-0.5234(z-0.7932)(z+0.009140)}{z^2 - z}.$$

The step responses for iterations 1 and 30 are presented in Fig. 3.9. Table 3.9 presents the numerators of the controller and the flexible reference model, obtained in different iterations. Even though Assumption B_{NMP} is not satisfied, the NMP zero is still identified with good precision by the minimization of $\tilde{J}_0^{VR}(\eta, \rho)$. Besides, the closed loop $T(z, \hat{\rho}_{30})$ presents a response that is not exactly, but very similar to the reference model $T_f(z, \hat{\eta}_{30})$ response (see Fig. 3.9).

We can again apply the second step of our procedure, modifying the fixed reference model to include the NMP zero just identified. The fixed reference model

Table 3.9 Evolution of $T_f(z,\eta)$, $C(z,\rho)$ (by means of their numerators—gains and zeros) and $\tilde{J}_0^{VR}(\hat{\eta},\hat{\rho})$ in the iterative procedure for the process $G_1(z)$

i	$\text{num}(T_f(z,\hat{\eta}_i))$	$\tilde{J}_0^{VR}(\hat{\eta}_i,\hat{\rho}_{i-1})$ ($\times 10^{-6}$)
1	$-0.9570(z-1.136)(z-0.5078)$	1.3176005
2	$-0.8313(z-1.160)(z-0.5205)$	4.6132843
10	$-0.6368(z-1.194)(z-0.4821)$	0.0731824
20	$-0.5768(z-1.196)(z-0.4349)$	0.0240805
30	$-0.5422(z-1.197)(z-0.4021)$	0.0108115

i	$\text{num}(C(z,\hat{\rho}_i))$	$\tilde{J}_0^{VR}(\hat{\eta}_i,\hat{\rho}_i)$ ($\times 10^{-6}$)
1	$-0.6904(z-0.7452)(z-0.2133)$	9.5254229
2	$-0.6659(z-0.7628)(z-0.1892)$	1.7675536
10	$-0.5905(z-0.7870)(z-0.09796)$	0.0683009
20	$-0.5491(z-0.7910)(z-0.03529)$	0.0231907
30	$-0.5234(z-0.7932)(z+0.009140)$	0.0105319

$T_{d_{f,m}}(z)$ should then be defined as

$$T_{d_{f,m}}(z) = \frac{-0.3242z(z-1.197)}{(z-0.6)^3}. \tag{3.63}$$

In Step 2, we find the following controller

$$C(z,\hat{\rho}) = \frac{-0.3350(z+0.5649)(z-0.8066)}{z^2-z}.$$

Figure 3.10 presents the reference models and the step responses obtained for Steps 1 and 2. Again, $T_{d_{f,m}}(z)$ allows the system to present a smaller inverse response, closer to the response specified by the original reference model $T_{d_f}(z)$.

Dealing with Noisy Data To show the applicability of the flexible reference model method to noisy systems, consider still the same process and control objective, and that the data are collected in closed-loop with the same controller (3.58), but now the output signal is affected by white noise, with variance $\sigma_e^2 = 0.005$. The optimal value of the controller parameter is already known, and can be used to assess the quality of the estimate obtained in this noisy scenario.

In order to assess the statistical properties of the design, we used 50 Monte Carlo runs to calculate the mean values and the variance of each element of the parameter vectors η and ρ. Initially, the algorithm was applied with LS optimization—without using instrumental variables. In this case, the mean value of $\hat{\eta}_{60}$, considering 60 iterations, has been obtained:

$$\hat{\eta}_{60,m} = \begin{bmatrix} -0.1986 \\ 0.08502 \\ 0.1841 \end{bmatrix}^T,$$

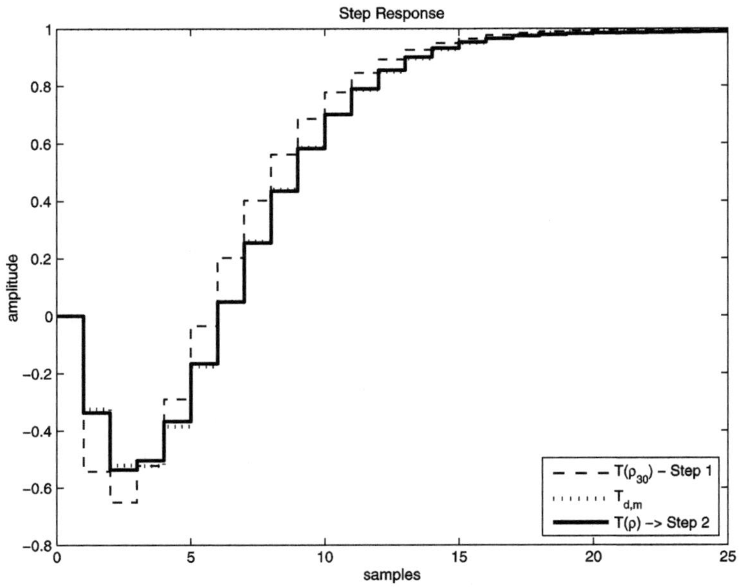

Fig. 3.10 Closed-loop responses $T(z, \hat{\rho})$ obtained at the end of Steps 1 and 2, compared to the modified reference model $T_{d_{f,m}}(z)$

which corresponds to a reference model

$$T(z, \hat{\eta}_{60,m}) = \frac{-0.1986(z - 1.201)(z + 0.7724)}{(z - 0.885)(z^2 - 0.706z + 0.32)}, \tag{3.64}$$

and the covariance of this estimate is given by

$$\text{var}_{\hat{\eta}} = 10^{-3} \times \begin{bmatrix} 1.984 & -4.244 & 2.260 \\ -4.244 & 9.109 & -4.865 \\ 2.260 & -4.865 & 2.605 \end{bmatrix}. \tag{3.65}$$

Also, the mean value of the corresponding $\hat{\rho}_{60}$ is

$$\hat{\rho}_{60,m} = \begin{bmatrix} -0.2950 \\ 0.1401 \\ 0.07875 \end{bmatrix}^T,$$

for which the corresponding controller is given by

$$C(z, \hat{\rho}_{60,m}) = \frac{-0.2950(z - 0.8060)(z + 0.3312)}{z^2 - z}. \tag{3.66}$$

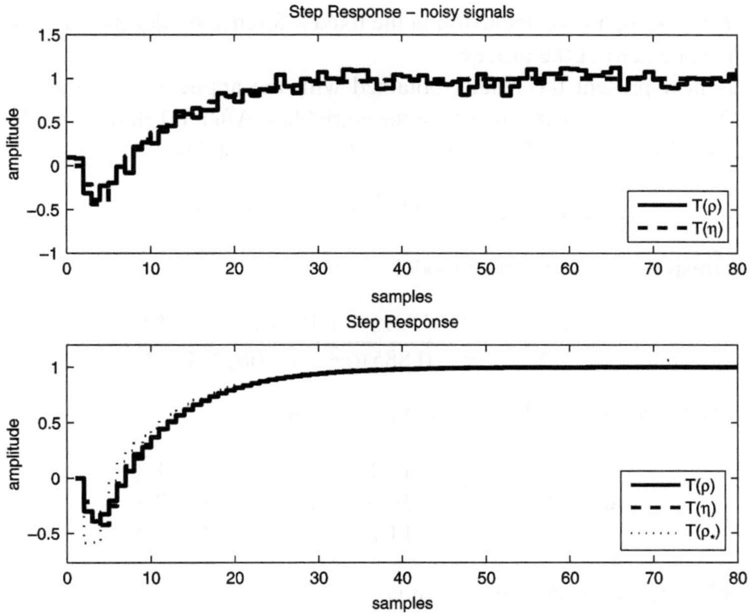

Fig. 3.11 Flexible reference model and closed-loop responses obtained applying the flexible VRFT criterion in a noisy system, without using instrumental variables

The covariance of the estimate is

$$\mathrm{var}_{\hat{\rho}} = 10^{-3} \times \begin{bmatrix} 0.9165 & -1.670 & 0.7680 \\ -1.670 & 3.060 & -1.416 \\ 0.7680 & -1.416 & 0.6594 \end{bmatrix}. \tag{3.67}$$

The controller parameters obtained in the noiseless case are given in Table 3.8. Comparing the average controller obtained in the presence of noise, described in (3.64) and (3.66), to the optimal values, a significant difference is observed—the results obtained with noisy data are biased. This difference can also be observed in Fig. 3.11, where the closed-loop response obtained with the mean controller (3.66) is compared to the response that would have been obtained with the mean flexible reference model (3.64) and with the "optimal" reference model $T(z, \hat{\rho}_{30})$, obtained previously. We show the noisy data in the top figure and we show the simulated result in a noiseless system in order to facilitate the visual assessment of the result. Note that $T(z, \hat{\rho}_{30}) \approx T(z, \hat{\eta}_{30})$, but it is still far from the optimal value $T(z, \eta_*)$.

Despite the results being biased, the non-minimum phase zero of the process is estimated with good precision, as well as the slower pole of the controller. One could think that the noise effect is not harmful, since the objective of this step is to identify the NMP-zero of the process and this was successfully done. However, the other zero of the reference model is far away from the optimal value, which is equal to the other zero of the process—0.4. In fact, this other zero could be estimated as another NMP zero, and its inclusion in the fixed reference model would be harmful, since it is not a zero of the process.

Let us now present the results obtained with the application of the algorithm (3.50)–(3.51), that is, using instrumental variables. After 60 iterations, the mean value (again 50 Monte Carlo runs were used) $\hat{\eta}_{60}$ obtained is

$$\hat{\eta}^{IV}_{60,m} = [\,-0.6192 \;\; 1.0034 \;\; -0.3137\,]^T,$$

which corresponds to a reference model

$$T(z, \hat{\eta}^{IV}_{60,m}) = \frac{-0.6192(z-1.198)(z-0.4230)}{(z-0.885)(z^2-0.706z+0.32)}. \tag{3.68}$$

The estimated variance of this estimate is given by

$$\mathrm{var}_{\hat{\eta}} = 10^{-3} \times \begin{bmatrix} 64.62 & -138.6 & 74.01 \\ -138.6 & 297.5 & -158.9 \\ 74.01 & -158.9 & 84.87 \end{bmatrix}. \tag{3.69}$$

Accordingly, the mean value of $\hat{\rho}_{60}$ is

$$\hat{\rho}^{IV}_{60,m} = [\,-0.5929 \;\; 0.6563 \;\; -0.1469\,]^T,$$

which corresponds to a controller given by

$$C(z, \hat{\rho}^{IV}_{60,m}) = \frac{-0.5929(z-0.7956)(z-0.3115)}{z^2-z}. \tag{3.70}$$

The estimated variance of this estimate is given by

$$\mathrm{var}_{\hat{\rho}} = 10^{-3} \times \begin{bmatrix} 20.78 & -36.27 & 16.16 \\ -36.27 & 63.40 & -28.30 \\ 16.16 & -28.30 & 12.66 \end{bmatrix}. \tag{3.71}$$

Now, comparing (3.68) and (3.70), obtained with instrumental variables, with the optimal values $T(z, \eta_*)$ and $C(z, \rho_*)$, we see that, in this case, the parameter's estimates are unbiased, and the results obtained using noisy data are very similar to the results obtained using noise-free data. The same conclusion can be extracted from the analysis of the results presented in Fig. 3.12.

3.6.2.2 Process with Two Minimum-Phase Zeros

Finally, we apply the method to an example in which the plant zeros are both minimum phase:

$$G_2(z) = \frac{(z+0.2)(z-0.4)}{z(z-0.3)(z-0.8)}. \tag{3.72}$$

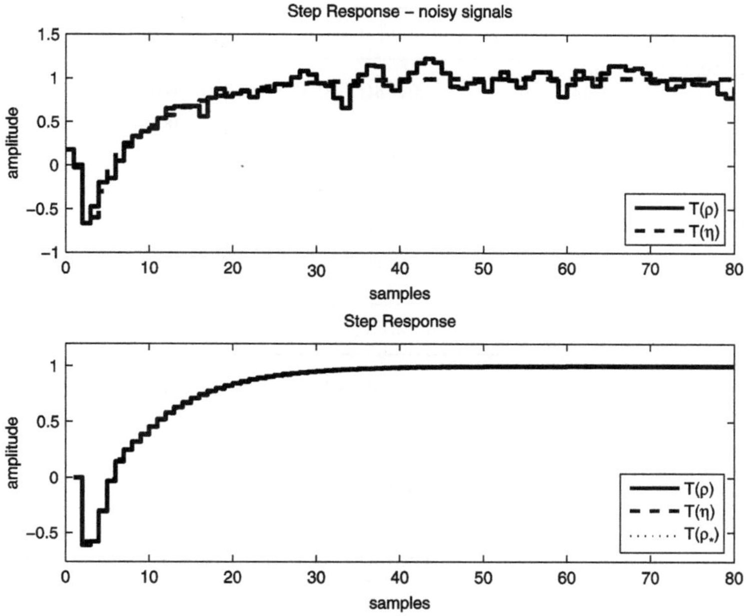

Fig. 3.12 Flexible reference model and closed-loop responses obtained applying the flexible VRFT criterion in a noisy system, using instrumental variables

This process is initially in closed loop with a PID controller

$$C_0(z) = \frac{0.7(z-0.4)(z-0.6)}{z^2 - z},$$

which we want to retune so that the closed-loop response is as close as possible to a given reference model, using a PID controller $C(z,\rho)$ of the form (3.57).

Assumption B_{NMP} Is Satisfied The desired fixed reference model is given by

$$T_d(z) = \frac{0.4601z^2}{(z-0.6673)(z^2+0.3063z+0.0766)},$$

and the flexible reference model is chosen as

$$T(z,\eta) = \frac{\vartheta_1 z^2 + \vartheta_2 z + \vartheta_3}{(z-0.6673)(z^2+0.3063z+0.0766)},$$

for which Assumption B_{NMP} is satisfied. After 40 iterations we obtain the following reference model and corresponding controller

$$T(z,\hat{\eta}_{40}) = \frac{0.6681(z-0.4157)(z+0.1785)}{(z-0.6673)(z^2+0.3063z+0.0766)},$$

Table 3.10 Evolution of $T_d(z,\eta)$, $C(z,\rho)$ (by means of their numerators—gains and zeros) and $\tilde{J}_0^{VR}(\hat{\eta},\hat{\rho})$ in the iterative procedure for the process $G_2(z)$

i	$\text{num}(T_d(z,\hat{\eta}_i))$	$\tilde{J}_0^{VR}(\hat{\eta}_i,\hat{\rho}_{i-1})$
1	$0.6788(z-0.4595)(z+0.2540)$	0.6958384
2	$0.6783(z-0.4566)(z+0.2482)$	0.0057458
5	$0.6768(z-0.4486)(z+0.2328)$	0.0036328
10	$0.6744(z-0.4377)(z+0.2134)$	0.0016586
30	$0.6691(z-0.4183)(z+0.1819)$	0.0001530
40	$0.6681(z-0.4157)(z+0.1785)$	0.0001157

i	$\text{num}(C(z,\hat{\rho}_i))$	$\tilde{J}_0^{VR}(\hat{\eta}_i,\hat{\rho}_i)$
1	$0.6776(z-0.8155)(z-0.2526)$	0.0401828
2	$0.6771(z-0.8146)(z-0.2568)$	0.0342863
5	$0.6756(z-0.8121)(z-0.2678)$	0.0211759
10	$0.6734(z-0.8087)(z-0.2817)$	0.0093376
30	$0.6687(z-0.8028)(z-0.3038)$	0.0010353
40	$0.6678(z-0.8020)(z-0.3063)$	0.0009334

$$C(z,\hat{\rho}_{40}) = \frac{0.6678(z-0.8020)(z-0.3063)}{z^2-z}.$$

In Step 1 the zeros of $T(z,\eta)$, estimated using (3.30)–(3.31), converge to the zeros of $G_2(z)$, but more slowly than in the case of NMP zeros: see Table 3.10. Since $T(z,\hat{\eta}_{40})$ does not have a NMP zero, we can safely go to Step 2 and use the standard VRFT method without modifying the reference model. The controller obtained with $T_d(z)$ is

$$C(z,\hat{\rho}) = \frac{0.4625(z-0.2996)(z-0.7613)}{z^2-z}.$$

Assumption B_{NMP} Is Not Satisfied Consider now the control of the same process under the same circumstances, but a different fixed reference model:

$$T_{d_f}(z) = \frac{0.216z^2}{(z-0.4)^3}, \tag{3.73}$$

and a flexible model having the same poles as $T_d(z)$:

$$T_f(z,\eta) = \frac{\vartheta_1 z^2 + \vartheta_2 z + \vartheta_3}{(z-0.4)^3}.$$

With $T_f(z,\eta)$ and the controller (3.57), Assumption B_{NMP} is not satisfied. Then Step 1 of the two steps procedure leads to

$$T_f(z,\hat{\eta}_{10}) = \frac{0.4986(z-0.6934)(z+0.4129)}{(z-0.4)^3},$$

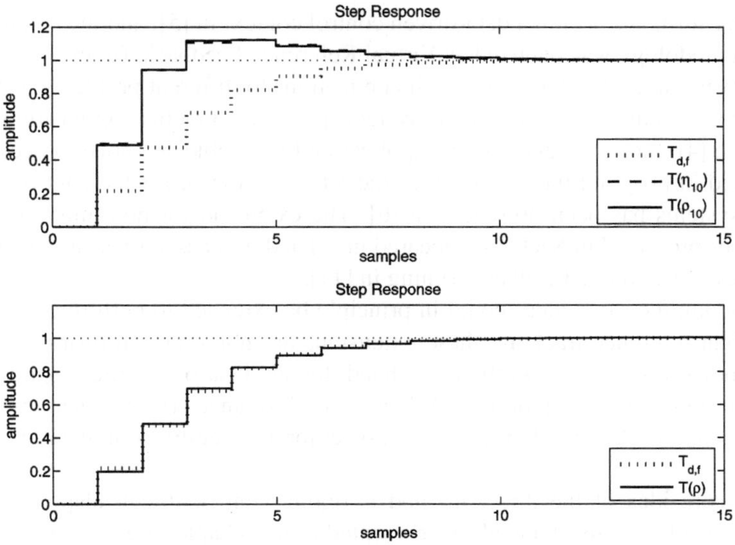

Fig. 3.13 Step responses of the fixed reference model $T_{d_f}(z)$, the flexible reference model $T_f(z, \hat{\eta}_{10})$ and the closed-loop response $T(z, \hat{\rho}_{10})$ obtained after 10 iterations in Step 1 (*top* figure); step responses of the original fixed reference model $T_{d_f}(z)$ and of the closed-loop system $T(z, \hat{\rho})$ obtained in Step 2 (*bottom*)

$$C_f(z, \hat{\rho}_{10}) = \frac{0.4956(z - 0.6815)(z + 0.2039)}{z^2 - z}.$$

Since the first step computes not only the numerator coefficients of the flexible reference model, but also a corresponding controller, one might consider applying this controller $C_f(z, \hat{\rho}_{10})$ to the plant, thereby avoiding the need for a second step. But the performance produced by the controller $C_f(z, \hat{\rho}_{10})$ is far from the desired performance specified by the original reference model $T_{d,f}(z)$, as illustrated in the top part of Fig. 3.13. We thus proceed to Step 2 and apply the standard VRFT with the fixed reference model $T_{d,f}(z)$. The controller thus obtained is

$$C(z, \hat{\rho}) = \frac{0.1931(z + 0.5997)(z - 0.7890)}{z^2 - z}.$$

The closed-loop response obtained with $C(z, \hat{\rho})$ is compared to the performance of the fixed reference model $T_{d_f}(z)$ at the bottom part of Fig. 3.13.

3.7 Chapter Conclusions

In this chapter we have presented a quadratic function whose global minimum is the same as the reference tracking criterion $J_y(\rho)$ under certain ideal conditions, and close to that under conditions close to this ideal. This idea was first presented in this clear form and within a data-driven control context in [5], and the resulting design methodology was baptized as Virtual Reference Feedback Tuning. There have been several later developments, with applications to different problems, including nonlinear systems and multivariable systems [4, 15]. A practical application is presented in [4], thoughts regarding its application to unstable and noisy processes are presented in [16] and the extension of the VRFT rationale for the control of nonlinear systems has been presented in [6]. The extension for non-minimum phase processes presented in Sect. 3.4 appeared in [3] and was based on a similar solution presented for Iterative Feedback Tuning in [11].

Although the same ideas could in principle be extended to performance criteria other than reference tracking, these extensions are not straightforward and, as of today, not yet available. On the other hand, the application of direct optimization relies strongly on Assumption B_y. When the performance criterion is not reference tracking, and/or Assumption B_y can not be enforced, iterative optimization is necessary.

The core idea of the direct method is reminiscent of direct adaptive control, and the developments and analyses presented in this chapter are closely related to, and strongly rely on, classical parameter identification theory. The classical textbooks on identification theory by Ljung [12] and by Söderström and Stoica [17] are paramount references here, as well as the ones on adaptive control by Åström and Wittenmark [1] and Goodwin and Sin [8]. The sufficient richness, persistence of excitation and transfer of excitation concepts, and the resulting conditions for uniqueness of the global minimum of prediction error and H_2 control performance criteria, are also amply studied in the identification and adaptive control literature. These are given within different contexts and with varying nomenclature and notation, and even under different sets of hypotheses; for an analysis and a full demonstration of uniqueness of the global minimum of prediction error criteria with the same notation, nomenclature and underlying assumptions presented in this book, see [7].

A more recent method for direct optimization of the reference tracking performance is the noniterative Correlation-based Tuning presented in [10]. This method is conceived from a different perspective, which is to minimize the correlation between the reference $r(t)$ and the tracking error $y(t,\rho) - y_d(t)$. Still, comparing to what has been presented in this chapter, the objective is the same (to minimize $J_y(\rho)$), the algorithmic aspects are quite similar and so are the main properties of the resulting estimate of the optimal controller parameter.

References

1. K.J. Åström, B. Wittenmark, *Adaptive Control*, 2nd edn. (Addison-Wesley, Reading, 1995)
2. R.R. Bitmead, S. Cheong, System identification based instability detection of noisy ARMA systems, in *Proc. 15th IFAC Symposium on System Identification*, Saint-Malo, France (IFAC, 2009), pp. 290–295
3. L. Campestrini, D. Eckhard, M. Gevers, A.S. Bazanella, Virtual reference feedback tuning for non-minimum phase plants. Automatica **47**, 1778–1784 (2011)

4. M.C. Campi, A. Lecchini, S. Savaresi, An application of the virtual reference feedback tuning method to a benchmark problem. Eur. J. Control **9**, 66–76 (2003)
5. M.C. Campi, A. Lecchini, S.M. Savaresi, Virtual reference feedback tuning: A direct method for the design of feedback controllers. Automatica **38**, 1337–1346 (2002)
6. M.C. Campi, S.M. Savaresi, Direct nonlinear control design: The virtual reference feedback tuning (VRFT) approach. Automatica **51**, 14–27 (2006)
7. M. Gevers, A.S. Bazanella, X. Bombois, L. Mišković, Identification and the information matrix: How to get just sufficiently rich? IEEE Trans. Autom. Control **54**(12), 2828–2840 (2009)
8. G. Goodwin, K.S. Sin, *Adaptive Filtering, Prediction and Control* (Prentice Hall International, Englewood Cliffs, 2009)
9. H. Hjalmarsson, M. Gevers, S. Gunnarsson, O. Lequin, Iterative feedback tuning: Theory and applications. IEEE Control Syst. Mag. **18**(4), 26–41 (1998)
10. A. Karimi, K. van Heusden, D. Bonvin, Non-iterative data-driven controller tuning using the correlation approach, in *Proc. European Control Conference*, Kos, Greece, 2007
11. A. Lecchini, M. Gevers, On iterative feedback tuning for non-minimum phase plants, in *41st IEEE Conference on Decision and Control*, 2002, pp. 4658–4663
12. L. Ljung, *System Identification—Theory for the User*, 2nd edn. (Prentice Hall, New York, 1999)
13. J. Mårtensson, Geometric Analysis of Stochastic Model Errors in System Identification, PhD thesis, Royal Institute of Technology (KTH), October 2007
14. J. Mårtensson, H. Hjalmarsson, Variance error quantifications for identified poles and zeros. Automatica **45**(11), 2512–2525 (2009)
15. M. Nakamoto, An application of the virtual reference feedback tuning for an MIMO process, in *Proc. SICE Annual Conference*, Sapporo, Japan, vol. 3 (IEEE Press, New York, 2005), pp. 2208–2213
16. A. Sala, A. Esparza, Extensions to "virtual reference feedback tuning: A direct method for the design of feedback controllers". Automatica **41**, 1473–1476 (2005)
17. T. Söderström, P. Stoica, *System Identification* (Prentice Hall, New York, 1989)

Chapter 4
Iterative Optimization

In this chapter we will study the optimization concepts that are central to the theory of data-driven control. This is by no means a tutorial on optimization; it is an assemblage of the topics and theoretical results immediately necessary to the general study of iterative data-driven control design.

In the next section, we recall some concepts from multivariable calculus. Some of these concepts are very basic ones, but we prefer to present them in our own way instead of leaving for the reader to remember them by him/herself. This serves to fix nomenclature and notation, and to call the reader's attention to those aspects of the basic optimization theory which are the most relevant for our purposes.

Then we present a convergence analysis of the fundamental optimization algorithms: steepest descent and Newton-Raphson. There are of course a myriad of other optimization algorithms to chose (Gauss-Newton, Levenberg-Marquardt, quasi-Newton, etc) but the main qualitative aspects of convergence analysis, which are the ones relevant for the analysis of iterative data-driven control design, are present in these two. We start, in Sect. 4.2, with a brief account of Lyapunov stability theory as applied to discrete-time systems. Then in Sect. 4.3 these stability analysis tools are applied to the analysis of convergence of the optimization algorithms. Last, but not least, we show that the properties and the conclusions presented are robust with respect to the unavoidable errors in the implementation of the algorithms.

4.1 Some Things to Remember from Calculus

Let $J(\cdot): \mathbb{R}^p \to \mathbb{R}^+$ be a function that is analytical in some domain $\mathcal{D} \subseteq \mathbb{R}^p$. The gradient of this function is a column vector annotated

$$\nabla J(\rho) \triangleq \frac{\partial J(\rho)}{\partial \rho}.$$

A. Sanfelice Bazanella et al., *Data-Driven Controller Design*,
Communications and Control Engineering,
DOI 10.1007/978-94-007-2300-9_4, © Springer Science+Business Media B.V. 2012

The Hessian of this function is a symmetric matrix annotated

$$\nabla^2 J(\rho) \triangleq \frac{\partial^2 J(\rho)}{\partial \rho^2}.$$

The Hessian is symmetric by construction and thus its eigenvalues are all real.

Definition 4.1 (Extrema) A point $\rho^1 \in \mathcal{D} \subseteq \mathbb{R}^p$ is said to be a *minimum* of $J(\cdot)$ if $\exists \epsilon > 0$ such that $J(\rho) \geq J(\rho^1)\ \forall \rho \in \mathcal{B}_\epsilon(\rho^1)$. It is said to be a *maximum* of $J(\cdot)$ if $\exists \epsilon > 0$ such that $J(\rho) \leq J(\rho^1)\ \forall \rho \in \mathcal{B}_\epsilon(\rho^1)$. It is said to be an *extremum* if it is either a minimum or a maximum.

A minimum ρ^1 is said to be[1]

- an *isolated minimum* if $\exists \epsilon > 0$ such that $J(\rho) > J(\rho^1)\ \forall \rho \in \mathcal{B}_\epsilon(\rho^1) \setminus \rho^1$;
- a *global minimum* in $\mathcal{D}$ if $J(\rho) \geq J(\rho^1)\ \forall \rho \in \mathcal{D}$;
- the *unique global minimum* in $\mathcal{D}$ if $J(\rho) > J(\rho^1)\ \forall \rho \in \mathcal{D} \setminus \rho^1$;
- a *local minimum* if it is a minimum but it is not a global minimum.

For ease of reference it is also convenient to define the following.

Definition 4.2 A given point ρ^1 is said to be a *critical point* of $J(\cdot)$ if $\nabla J(\rho^1) = \mathbf{0}$.

The main motivation to give a name to the points at which the gradient vanishes is the following fact.

Lemma 4.1 *Any extremum is a critical point.*

Proof Let ρ^1 be not a critical point of $J(\cdot)$—that is, $\nabla J(\rho^1) \neq \mathbf{0}$. We will prove that such a ρ^1 can not be an extremum.

Let $\mathbf{d}$ be an unitary vector such that $\mathbf{d}^T \nabla J(\rho^1) \neq 0$. Because $J(\cdot)$ is analytical and $\mathbf{d}$ is unitary, we can write

$$J(\rho^1 + \delta \mathbf{d}) - J(\rho^1) = \delta \nabla J(\rho^1)\mathbf{d} + \text{hot}(\delta)$$

for any $\delta \in \mathbb{R}$, where hot(δ) stands for "higher order terms", in this case terms of order higher than one, meaning that

$$\lim_{\delta \to 0} \frac{\text{hot}(\delta)}{\delta} = 0.$$

Thus, $\exists \delta_0$ such that $\|\delta \nabla J(\rho^1)\mathbf{d}\| > \|\text{hot}(\delta)\|$ for all δ such that $|\delta| < \delta_0$. Hence, for any $|\delta| < \delta_0$, the sign of $J(\rho^1 + \delta\mathbf{d}) - J(\rho^1)$ is determined by the product $\nabla J(\rho^1)\delta\mathbf{d}$ alone:

$$\text{sign}[J(\rho^1 + \delta \mathbf{d}) - J(\rho^1)] = \text{sign}[\nabla J(\rho^1)\delta \mathbf{d}] \quad \forall \delta : |\delta| < \delta_0.$$

[1] Of course similar definitions can be made for maxima.

It is thus clear that, as δ crosses through zero, $J(\rho^1 + \delta\mathbf{d}) - J(\rho^1)$ changes sign. So there are points arbitrarily close to ρ^1 for which $J(\rho) > J(\rho^1)$ (which implies that ρ^1 is not a maximum) and others for which $J(\rho) < J(\rho^1)$ (which implies that ρ^1 is not a minimum). Hence, ρ^1 is not an extremum. □

It is important to notice that this is a necessary but not sufficient condition, that is, not all critical points are extrema. Besides the extrema, inflexion points are also critical points.

Lemma 4.1 establishes that all minima and maxima share a property: the gradient vanishes at these points. It remains to establish what differentiates maxima from minima, which is done by analyzing the next derivative of $J(\cdot)$.

Lemma 4.2 *Let ρ^1 be a critical point of $J(\cdot)$. Then the following facts hold.*

- *ρ^1 is an isolated minimum of $J(\cdot)$ if $\nabla^2 J(\rho^1) > 0$;*
- *ρ^1 is an isolated maximum of $J(\cdot)$ if $\nabla^2 J(\rho^1) < 0$.*

Proof We will prove the first item of the statement; the proof for the other item is the same, *mutatis mutandis*.

Since $\nabla J(\rho^1) = \mathbf{0}$ we can write

$$J(\rho^1 + \delta\mathbf{d}) - J(\rho^1) = \frac{1}{2}\delta^2 \mathbf{d}^T \nabla^2 J(\rho^1)\mathbf{d} + \text{hot}(\delta) \tag{4.1}$$

with $\mathbf{d}$ an arbitrary unitary vector and hot(δ) now satisfying

$$\lim_{\delta \to 0} \frac{\text{hot}(\delta)}{\delta^2} = 0.$$

By the same argument in Lemma 4.1, the sign of $J(\rho^1 + \delta\mathbf{d}) - J(\rho^1)$ is determined exclusively by the first term in the right hand side of (4.1) for sufficiently small δ, that is:

$$\text{sign}[J(\rho^1 + \delta\mathbf{d}) - J(\rho^1)] = \text{sign}[\mathbf{d}^T \nabla^2 J(\rho^1)\mathbf{d}] \quad \forall\delta : |\delta| < \delta_0 \tag{4.2}$$

for some $\delta_0 > 0$ and for all unitary $\mathbf{d}$. If the Hessian is positive definite, then the right hand side of (4.2) is always positive, and thus $J(\rho^1 + \delta\mathbf{d}) > J(\rho^1)$ $\forall\delta : |\delta| < \delta_0$; that is, ρ^1 is an isolated minimum. □

If the Hessian is not positive definite, then at least one of its eigenvalues is negative or zero[2]. Letting $\mathbf{v}$ be an eigenvector associated to a negative eigenvalue, we have $J(\rho^1 + \delta\mathbf{v}) \leq J(\rho^1)$ $\forall\delta : |\delta| < \delta_0$, that is, there are points arbitrarily close to ρ^1 for which the function's value is smaller than its value at ρ^1; hence ρ^1 is not a minimum.

---

[2]Note that the Hessian is symmetric and thus all its eigenvalues are real.

An obvious fact deriving from the above lemma is that an analytical function always presents a convex behavior in the neighborhood of an isolated extremum. Let us define convexity as follows.

Definition 4.3 A function $J(\cdot)$ is said to be convex in a domain Ω if $\nabla^2 J(\rho) \geq 0$ $\forall \rho \in \Omega$ and then the set Ω is called a *domain of convexity*. It is said to be concave in Ω if $\nabla^2 J(\rho) \leq 0$ $\forall \rho \in \Omega$ and then the set Ω is called a *domain of concavity*.

Then, with this nomenclature, it is clear from the very definition of extrema that there exists a nonempty domain of convexity around any isolated minimum of an analytical function $J(\cdot)$. Similarly, there exists a nonempty domain of concavity around any isolated maximum of an analytical function $J(\cdot)$. This evident and almost tautological fact is not always properly appreciated in the applications of optimization.

These properties relating a function's extrema and its derivatives inspire the derivation of many optimization algorithms and form the foundations for their analysis, as will be seen next.

4.2 Optimization Algorithms (Dynamic Systems)

An iterative optimization algorithm is a recursion, which generates successive approximations for the global minimum of $J(\rho)$:

$$\rho_{i+1} = f(\rho_i, i) \tag{4.3}$$

where $f(\cdot,\cdot) : \mathbb{R}^p \times \mathbb{Z}^+ \to \mathbb{R}^p$. Notice that at each iteration the function $f(\cdot,\cdot)$ is evaluated at the current approximation ρ_i, that is, only this "local" information (information at the current approximation) is used in the optimization. If the algorithm is to be of any use, then the function $f(\rho_i, i)$ should be such that, as the iterations pass, the approximations ρ_i approach a minimum, eventually converging to it. And preferentially, this minimum to which the algorithm converges should be a global minimum.

Equation (4.3) defining the algorithm is a discrete-time dynamic system, where ρ_i is the system's state at "time" (iteration) i. It thus seems very reasonable to profit of systems' stability theory to study the convergence of these algorithms, since this is a classical and well established discipline. The convergence problem is approached by viewing the algorithms as nonlinear dynamical systems, and then analyzing their convergence as the stability of the corresponding system. Let us thus follow with a brief review of nonlinear systems theory.

Definition 4.4 (Fixed-point) A vector ρ^1 is called a fixed-point (or an equilibrium) of an algorithm (4.3) if $f(\rho^1, i) = \rho^1$ $\forall i$. A fixed-point is said to be attractive (or asymptotically stable) if initializations sufficiently close to the fixed-point converge to it, that is, if there is a positive scalar ϵ such that for all $\rho_0 \in \mathcal{B}_\epsilon(\rho^1)$, $\lim_{i\to\infty} \rho_i = \rho^1$.

The name "fixed-point" expresses the fact that once the algorithm has reached this point, it can never move away from it or, in more mathematical terms:

$$\rho_k = \rho^1 \rightarrow \rho_i = \rho^1 \quad \forall i > k.$$

Whether fixed points are attractive or not is perhaps the main subject of nonlinear systems analysis. This problem can be tackled by the two Lyapunov methods: the direct and the indirect method. The indirect method involves the linearization of the nonlinear system (4.3) around the fixed point and its application to the analysis of the convergence of algorithms is commonplace in the literature. This analysis is useful to determine whether or not a fixed point is attractive, that is, whether or not initializations of the algorithm sufficiently close to the fixed point converge to it. But it is also important that convergence is observed for a large enough set of initializations of the algorithm. This analysis is not covered by the indirect method and will be our next concern.

Definition 4.5 Let ρ^1 be an attractive fixed-point of an algorithm $\rho_{i+1} = f(\rho_i, i)$. A p-dimensional set $\Omega \subset \mathbb{R}^p$ is a domain of attraction (DOA) of ρ^1 if $\lim_{i\to\infty} \rho_i = \rho^1 \; \forall \rho_0 \in \Omega$.

Notice that Ω in this definition is NOT the set of ALL initializations that converge to the fixed-point. This last set can be a very complex subset of the state space. As such, the set of all initializations that converge to the fixed point is in general very hard, if not impossible, to characterize in any meaningful way. On the other hand, it is not critical in the context of data-driven control design to know all the initializations that converge to the fixed-point, but only if these initializations form a "big enough" set, so that convergence to the fixed-point is obtained even when the optimization is initialized "far" from it. Accordingly, the analysis will focus in finding sets that satisfy Definition 4.5 and are also relatively easy to characterize mathematically. The standard sets are the ellipsoids, which can be described by equations like

$$\Omega = \{\rho \, : \, (\rho - \rho^1)^T P(\rho - \rho^1) < 1\} \tag{4.4}$$

where P is a symmetric positive definite matrix. The ellipsoid Ω described in (4.4) has ρ^1 as its center, its size is determined by the eigenvalues of P, and its shape by the eigenvectors of P. This set particularizes to a hypersphere (which we call a "ball" in this book) when $P = aI_n$ for some positive scalar a. Equations like (4.4) are very amenable to calculations and theoretical manipulations, so that it is not by chance that ellipsoids are the favorite sets to work with.

To analyze domains of attraction it is necessary to resort to the direct method of Lyapunov, which will be presented next. For this presentation, we still need an additional preparatory concept.

Definition 4.6 A function $\phi(\cdot) : \mathbb{R}^+ \rightarrow \mathbb{R}^+$ is of *class K* if it is continuous, strictly increasing and $\phi(0) = 0$.

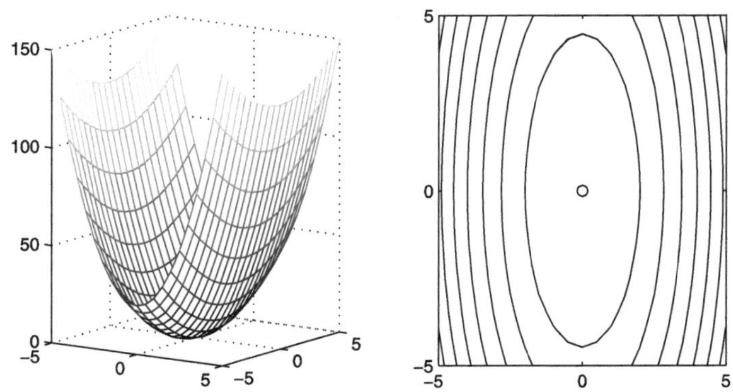

Fig. 4.1 A quadratic Lyapunov function $V(\rho) = \rho^T P\rho$, with P a positive definite matrix, and its level sets, which are ellipses in the form (4.4)

Now we are ready to state the basic Theorem of Lyapunov's direct method.

Theorem 4.1 (Lyapunov's direct method [6]) *Consider an algorithm* (4.3) *with a fixed point* ρ^1. *The fixed point* ρ^1 *is asymptotically stable if there exist a function* $V(\cdot) : \mathbb{R}^p \rightarrow \mathbb{R}^+$, *class K functions* $\alpha(\cdot)$, $\beta(\cdot)$ *and* $\omega(\cdot)$, *and a positive scalar* ϵ *such that*:

$$\alpha(\|\rho - \rho^1\|) \leq V(\rho) \leq \beta(\|\rho - \rho^1\|) \quad \forall \rho \in \mathcal{B}_\epsilon(\rho^1) \tag{4.5}$$

$$V(f(\rho, i)) - V(\rho) \leq -\omega(\|\rho - \rho^1\|) \quad \forall \rho \in \mathcal{B}_\epsilon(\rho^1), \forall i. \tag{4.6}$$

When a function $V(\cdot)$ satisfying the conditions of the theorem is found, then it is called a Lyapunov function, and the asymptotic stability of the fixed point is established. The main feature of the direct method, when compared to the indirect method, is that it allows to determine DOAs: level sets of a Lyapunov function serve as DOAs. The level sets are defined below.

Definition 4.7 (Level sets) The level sets of a Lyapunov function $V(\cdot)$ are defined as $\mathcal{L}_c \triangleq \{\rho : V(\rho) \leq c\}$ for scalars $c > 0$. Notice that by construction, $\rho_* \in \mathcal{L}_c\ \forall c$ (because by definition $V(\rho_*) = 0 < c$).

Figure 4.1 illustrates a typical Lyapunov function and its level sets. A level set $\mathcal{L}_c$ is a DOA of an asymptotically stable fixed-point ρ^1 if $\mathcal{L}_c \subset \mathcal{B}_\epsilon(\rho^1)$, where $\mathcal{B}_\epsilon(\rho^1)$ is the ball in the statement of Theorem 4.1. For a given Lyapunov function, the largest c for which this inclusion is satisfied provides the best (largest) DOA that can be obtained with this Lyapunov function. In the next section, these concepts on the stability of nonlinear systems will be applied to the analysis of fundamental optimization algorithms.

4.2.1 Autonomous Systems

It will often be the case that the recursion defining the algorithm can be written in the form

$$\rho_{i+1} = f(\rho_i). \tag{4.7}$$

That is, the new parameter value ρ_{i+1} does not depend explicitly and independently on the "time" variable i as in (4.3), but only on the current state ρ_i. Nonlinear systems in this form are called *autonomous* and constitute a particular class of major relevance. All the results presented before are also valid for this class of systems, of course, but the application of the direct method in this case is simpler. The following Theorem particularizes the direct method of Lyapunov for autonomous systems.

Theorem 4.2 (Direct method—autonomous systems) *Consider an algorithm described by* (4.7) *with a fixed point* ρ^1. *The fixed point* ρ^1 *is asymptotically stable if there exist a function* $V(\cdot): \mathbb{R}^p \to \mathbb{R}^+$ *and a positive scalar* ϵ *such that*:

$$V(\rho) > 0 \quad \forall \rho \in \mathcal{B}_\epsilon(\rho^1) \tag{4.8}$$

$$V(f(\rho)) - V(\rho) < 0 \quad \forall \rho \in \mathcal{B}_\epsilon(\rho^1). \tag{4.9}$$

4.3 The Basic Algorithms and Their Convergence

What information is available at step i of the optimization? Knowing ρ_i it is usually possible to compute information about the function $J(\cdot)$ at this value of its argument: its value, its derivative, its Hessian. So, typical optimization algorithms make use of these informations and can be generically described by

$$f(\rho_i, i) = \rho_i - \gamma_i R_i \nabla J(\rho_i) \tag{4.10}$$

where, at each iteration i, R_i is a matrix and γ_i is a positive scalar. The matrices R_i define the direction of the parameter update made at step i, whereas the scalars γ_i define the magnitude of this update at each step. Each different policy for choosing the update directions R_i implies different convergence and implementation properties for the algorithm and is accordingly baptized: steepest descent and Newton-Raphson are the fundamental choices. Let us analyze each one of these algorithms separately.

4.3.1 Steepest Descent

The most basic optimization algorithm is the steepest descent. A steepest descent algorithm is one in which $R_i = I_p$ in (4.10), that is, the iteration is given by

$$\rho_{i+1} = \rho_i - \gamma_i \nabla J(\rho_i) \tag{4.11}$$

with $\gamma_i > 0 \; \forall i$. The rationale behind this algorithm is clear: updates are made in the opposite direction of the gradient, so, at least for sufficiently small γ_i, at each iteration a smaller value for the cost is obtained. Its properties are analyzed in the sequel, starting with the characterization of its fixed-points. The following lemma is an obvious consequence of the definitions of fixed-point and critical point, and also of (4.11).

Lemma 4.3 *A given point ρ^1 in the parameter space is a fixed-point of the steepest descent algorithm if and only if it is a critical point of the objective function $J(\rho)$.*

So, every minimum, maximum or inflexion point is a fixed-point of the steepest descent algorithm. But which fixed-points are asymptotically stable? Stability depends on the choice of the step sizes γ_i, but only the minima can be asymptotically stable for $\gamma_i > 0$. Indeed, taking as a Lyapunov function $V(\rho) = J(\rho) - J(\rho^1)$, where ρ^1 is a minimum, it clearly satisfies the first condition of Theorem 4.1. Satisfaction of the second condition depends on the γ_i's: they must be sufficiently small to guarantee that the function decreases at every iteration, yet large enough to guarantee that this decrease is bounded by a class K function.[3] By a similar argument, the maxima and inflexion points of the function $J(\cdot)$ are unstable fixed-points of the steepest descent algorithm, which is a good thing—the algorithm converges only to minima.

A formal statement of the convergence to the global minimum will be given shortly, but using a different Lyapunov function. First, we present an example illustrating these ideas.

Example 4.1 Consider the following objective function

$$J(\rho) = e^{\rho}(\rho - 1)^2$$

which is plotted in Fig. 4.2. This objective function has two finite extrema at $\rho = -1$ and $\rho = 1$. The point $\rho = 1$ is a minimum and $\rho = -1$ is a maximum of the function. Let us apply the steepest descent algorithm (4.11) to seek the minimum of the function.

The gradient of the cost function is given by

$$\nabla J(\rho) = e^{\rho}(2\rho - 2) + e^{\rho}(\rho - 1)^2.$$

Following the previous discussion, there are essentially two parameters that determine the convergence of the steepest descent algorithm: the initial condition and the step size. Three initial conditions were tried, in order to illustrate the convergence properties in this example, with different step size policies.

The first initial condition is $\rho_0 = -0.5$. Let us try a constant step size: $\gamma_i = 0.2\ \forall i$. The evolution of the algorithm for 20 iterations is presented in Fig. 4.3, where it can be seen that the algorithm converges to the global minimum $\rho = 1$.

---

[3] This second constraint is not present when the algorithm is an autonomous system.

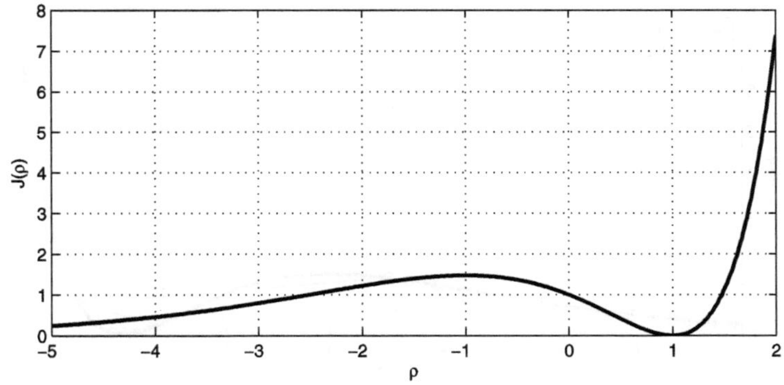

Fig. 4.2 Cost function $J(\rho) = e^{\rho}(\rho - 1)^2$

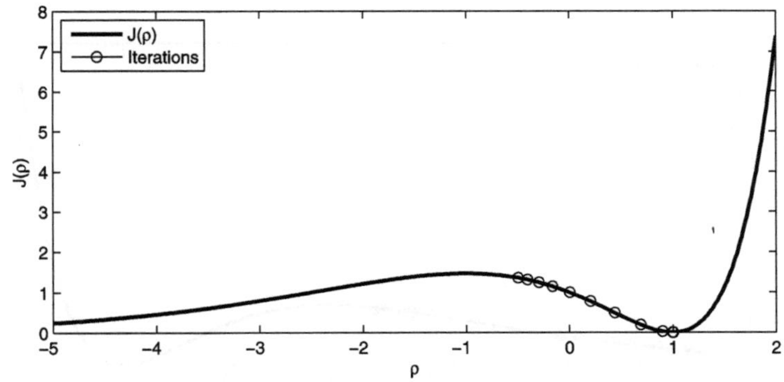

Fig. 4.3 Convergence of the steepest descent algorithm with initial condition $\rho_0 = -0.5$ and step size $\gamma_i = 0.2\ \forall i$

For the steepest descent algorithm with these step sizes the global minimum is an attractive fixed-point, and the initial condition $\rho_0 = -0.5$ is within a DOA of the global minimum.

As a second choice, let us try the initial condition $\rho_0 = 1.1$ with step sizes $\gamma_i = 0.7\ \forall i$. The resulting behavior of the algorithm for 20 iterations can be seen in Fig. 4.4. The algorithm diverged away from the global minimum $\rho = 1$, even though the initial condition was very closed to it, because of an unfortunate choice of the step sizes. With these step sizes the global minimum is not an attractive fixed-point.

The third initial condition is $\rho_0 = -1.5$, $\gamma_i = 0.5\ \forall i$. This again results in divergence, as shown in Fig. 4.5, but now this is because the initial condition was too far away from the global minimum. Even though the fixed-point is attractive with this choice of step sizes, this new initial condition does not belong to a DOA.

We have established theoretically and illustrated by an example that the steepest descent algorithm converges to a minimum provided that an appropriate step se-

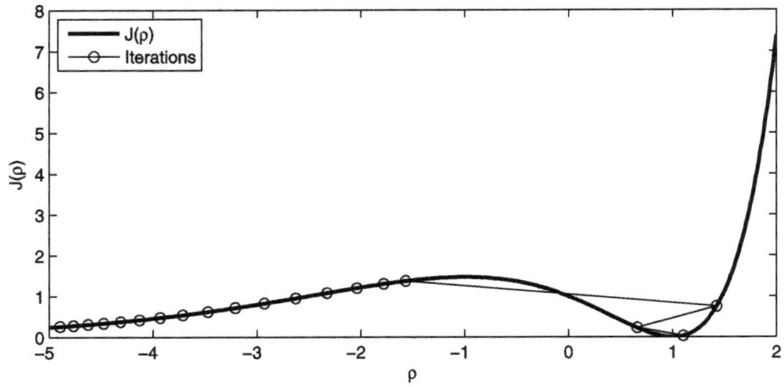

Fig. 4.4 Divergence of the steepest descent algorithm with initial condition $\rho_0 = 1.1$ and step size $\gamma_i = 0.7\ \forall i$

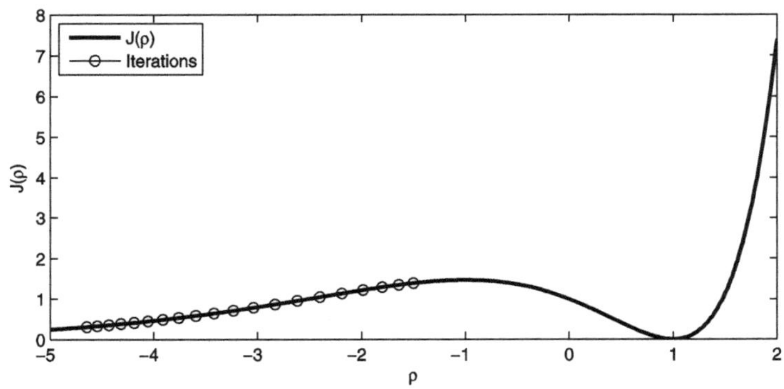

Fig. 4.5 Divergence of the steepest descent algorithm with initial condition $\rho_0 = -1.5$ and step size $\gamma_i = 0.5\ \forall i$

quence is chosen. This is nice, but we are still a long way from a full understanding of the convergence properties that are relevant in our applications. First, choosing the appropriate step sequence is far from trivial. Second, we would like the algorithm to converge to the global minimum, not just any minimum. And third, actual convergence will only be observed in practice if there exists a large enough DOA. The following result advances in these aspects and can also be generalized for other search directions besides the steepest descent.

Theorem 4.3 *Let ρ_* be an isolated global minimum of $J(\cdot) : \mathbb{R}^p \to \mathbb{R}^+$. Define the set $\mathcal{B}_\alpha(\rho_*) = \{\rho : (\rho - \rho_*)^T(\rho - \rho_*) < \alpha\}$ and a class K function $\omega(\cdot)$. If*

$$(\rho - \rho_*)^T \nabla J(\rho) > \|\nabla J(\rho)\| \sqrt{\omega(\|\rho - \rho_*\|)} > 0 \quad \forall \rho \in \mathcal{B}_\alpha(\rho_*) \setminus \rho_* \tag{4.12}$$

then there exists a sequence γ_i, $i = 1, \dots, \infty$ such that ρ_* is asymptotically stable and $\mathcal{B}_\alpha(\rho_*)$ is a DOA of algorithm (4.11) *for $J(\rho)$.*

Proof Let $V(\rho) = (\rho - \rho_*)^T(\rho - \rho_*)$ be a candidate Lyapunov function for the discrete-time system (4.11). Then

$$\begin{aligned}
&V(\rho_{i+1}) - V(\rho_i) + \omega(\|\rho - \rho_*\|) \\
&= (\rho_i - \gamma_i \nabla J(\rho_i) - \rho_*)^T(\rho_i - \gamma_i \nabla J(\rho_i) - \rho_*) - (\rho_i - \rho_*)^T(\rho_i - \rho_*) \\
&= -2\gamma_i(\rho_i - \rho_*)^T \nabla J(\rho_i) + \gamma_i^2 \nabla J(\rho_i)^T \nabla J(\rho_i) + \omega(\|\rho - \rho_*\|)
\end{aligned}$$

which is negative provided that

$$\begin{aligned}
\gamma_i &> \frac{(\rho_i - \rho_*)^T \nabla J(\rho_i) - \sqrt{((\rho_i - \rho_*)^T \nabla J(\rho_i))^2 - \nabla J(\rho_i)^T \nabla J(\rho_i)\omega(\|\rho_i - \rho_*\|)}}{\nabla J(\rho_i)^T \nabla J(\rho_i)} \\
\gamma_i &< \frac{(\rho_i - \rho_*)^T \nabla J(\rho_i) + \sqrt{((\rho_i - \rho_*)^T \nabla J(\rho_i))^2 - \nabla J(\rho_i)^T \nabla J(\rho_i)\omega(\|\rho_i - \rho_*\|)}}{\nabla J(\rho_i)^T \nabla J(\rho_i)}.
\end{aligned} \tag{4.13}$$

For $\rho_i \in \mathcal{B}_\alpha(\rho_*)$ the existence of such γ_i is guaranteed by condition (4.12), which also implies that $\nabla J(\rho_i) \neq \mathbf{0}\ \forall \rho_i \in \mathcal{B}_\alpha(\rho_*) \setminus \rho_*$. The proof is completed by noting that $\mathcal{B}_\alpha(\rho_*)$ is a connected and bounded level set of $V(\rho)$. □

The proof of Theorem 4.3 is constructive: it gives an explicit condition for the appropriate γ_i in (4.13), which in principle allows the appropriate choice of a step size sequence. But the γs in this equation depend on the global minimum itself, so this condition can not in principle be enforced, or even verified. In fact, with some more analytical effort it can, at least in data-driven control, as will be shown in Chap. 5. On the other hand, the calculation of the right hand side of (4.13) also requires the specification of the class K function $\omega(\cdot)$. This can be any class K function, and several choices can be made, resulting in different step size sequences. A simple, and thus natural, choice is $\omega(\|\rho - \rho_*\|) = a^2 \|\rho - \rho_*\|^2$ with $a^2 < 1$. With this choice, the inequality in condition (4.12) becomes:

$$(\rho - \rho_*)^T \nabla J(\rho) > a \|\nabla J(\rho)\| \, \|\rho - \rho_*\| .$$

But

$$(\rho - \rho_*)^T \nabla J(\rho) = \cos(\theta) \|\nabla J(\rho)\| \, \|\rho - \rho_*\| ,$$

where θ is the angle between the vectors $(\rho - \rho_*)$ and $\nabla J(\rho)$. Hence, condition (4.12) can be written as

$$\cos(\theta) \triangleq \frac{(\rho - \rho_*)^T \nabla J(\rho)}{\|\nabla J(\rho)\| \, \|\rho_i - \rho_*\|} > a \quad \forall \rho \in \mathcal{B}_\alpha(\rho_*) \setminus \rho_*. \tag{4.14}$$

Condition (4.14) is actually quite intuitive, if interpreted geometrically. Each step of the steepest descent algorithm is taken in the opposite direction of the gradient. Condition (4.14) means that at each point of the set $\mathcal{B}_\alpha(\rho_*)$ the component of the gradient along the direction $\rho - \rho_*$ points away from the global minimum ρ_*. Hence

Fig. 4.6 Gradient direction

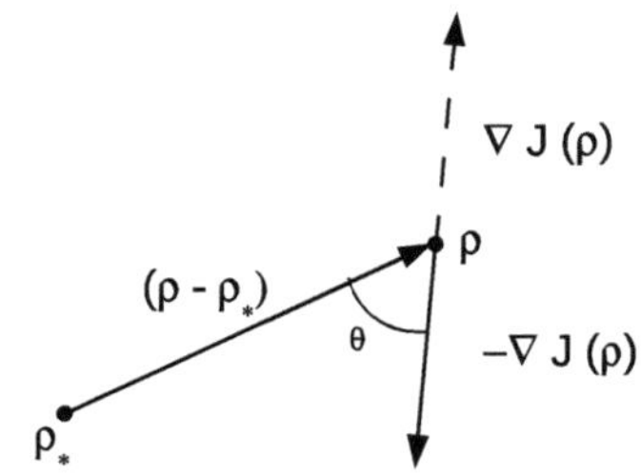

its negative, which is the direction taken by the algorithm, points towards it—see Fig. 4.6. Should this condition not be satisfied at some point of the set $\mathcal{B}_\alpha(\rho_*)$, then from this point the algorithm would take a step in the wrong direction, moving away from the global minimum, and then convergence to the global minimum can not be guaranteed if the algorithm is started at this point. Also notice that condition (4.12) implies that the gradient is never zero—there can be neither inflexion points nor other extrema in $\mathcal{B}_\alpha(\rho_*)$.

A set satisfying the conditions of Theorem 4.3 is such that it can be turned into a DOA by using the appropriate step sizes γ_i. Such a set is, so to say, a candidate to be a DOA. Accordingly, we define the following to facilitate further references to this property.

Definition 4.8 A ball $\mathcal{B}_\alpha(\rho_*)$ satisfying the conditions of Theorem 4.3 is called a *candidate DOA*.

4.3.2 Other Search Directions

The results just presented for the steepest descent algorithm are easily extended for other search directions, that is, for other matrices R_i in (4.10). We spare the reader the detailed repetition of the equations and Theorems, and focus on the main conceptual aspect involved in the use of a different search direction. The main condition for convergence to the global minimum in the general case of arbitrary matrices R_i will be, as in the steepest descent method, that the search direction "points" towards the global minimum and not away from it. In the particular case of the steepest descent the search direction is the opposite of the gradient and the exact expression of this condition is given in (4.14). The corresponding condition here will be the following

$$\cos(\theta_i) > a \quad \forall i \tag{4.15}$$

for some positive number $a < 1$, where θ_i is the angle between the vector $\rho_i - \rho_*$ and the search direction $R_i \nabla J(\rho_i)$, defined as

$$\cos(\theta_i) \triangleq \frac{(\rho_i - \rho_*)^T R_i \nabla J(\rho_i)}{\|R_i \nabla J(\rho_i)\| \, \|\rho_i - \rho_*\|}.$$

Observe that, unlike (4.14), this is not an *a priori* condition, that is, a condition that depends only on the properties of the cost function. This is because the search direction does not depend only on the cost function anymore, as was the case in the steepest descent method, but also on the (in principle arbitrary) matrices R_i at each iteration.

4.3.3 Newton-Raphson

Another fundamental optimization algorithm is the Newton-Raphson algorithm, which is defined by the following recursion:

$$\rho_{i+1} = \rho_i - (\nabla^2 J(\rho_i))^{-1} \nabla J(\rho_i). \tag{4.16}$$

In this recursion defining the Newton-Raphson algorithm, the search direction is no longer an arbitrary choice that varies with the iteration i, and neither is the step size. Instead, the search direction is defined as a function of the parameter vector ρ_i and the step sizes are fixed at $\gamma_i \equiv 1$. As a result, (4.16) defines an autonomous nonlinear system.

Unlike the steepest descent method, which was conceived with optimization in mind, the Newton-Raphson's method was conceived as a means to find the roots of a function. What the Newton-Raphson algorithm really does is to search for the roots of the equation $\nabla J(\rho) = \mathbf{0}$, which, by Theorem 4.1, are also extrema of $J(\rho)$. As a consequence, we can not expect this algorithm to tell minima from maxima, and, as will be seen shortly, this has undesired consequences.

Also notice that the algorithm involves inversion of the Hessian. When the function has several extrema, the Hessian will be singular at some points, and at these points the algorithm is not defined. Let us check the properties of the Newton-Raphson algorithm as we did for the steepest descent, excluding these singularity points from the analysis. The following is a direct consequence of the previous definitions.

Lemma 4.4 *Let ρ^1 be such that the Hessian $\nabla^2 J(\rho^1)$ is nonsingular, but otherwise arbitrary. Such a ρ^1 is a fixed-point of the Newton-Raphson algorithm* (4.16) *if and only if it is a critical point of the objective function $J(\rho)$.*

Concerning the stability of the fixed-points, we have the following property.

Theorem 4.4 *Let ρ^1 be an isolated extremum of a function $J(\rho)$. Then ρ^1 is an attractive fixed-point of the Newton-Raphson algorithm.*

Proof (sketch)[4] Let ρ^1 be a minimum, define $V(\rho) \triangleq J(\rho) - J(\rho^1)$ and let $\Omega \ni \rho^1$ be a domain of convexity around it. Then by definition $V(\rho^1) = 0$ and $V(\rho) > 0$

[4]For a more complete proof the reader is referred to standard optimization books. Such proofs get somewhat technical, so we prefer to give here only a sketch that provides insight into the convergence mechanisms.

$\forall \rho \in \Omega$. If $V(\rho_i)$ is strictly decreasing in time for all $\rho_i \in \Omega$, this makes it a Lyapunov function and proves the theorem.

At step i the Newton-Raphson algorithm gives: $\rho_{i+1} - \rho_i = -(\nabla^2 J(\rho_i))^{-1} \times \nabla J(\rho_i)$. Multiplying this expression on both sides by the gradient yields:

$$\nabla V(\rho_i)^T(\rho_{i+1} - \rho_i) = -\nabla J(\rho_i)^T(\nabla^2 J(\rho_i))^{-1}\nabla J(\rho_i) < 0 \quad \forall \rho_i \in \Omega \tag{4.17}$$

because $\nabla V(\rho_i) = \nabla J(\rho_i)$ and the Hessian is positive definite in Ω. But the expression in the left hand side of (4.17) is proportional to the directional derivative of $V(\rho)$ along the direction $\rho_{i+1} - \rho_i$. Because this is negative, so is the directional derivative and thus the function $V(\rho_i)$ is reduced when ρ is displaced in this direction, at least for sufficiently small displacements. Because the gradient tends to zero as the minimum is approached, so does the size of the displacements, and thus there exists a neighborhood around it for which these displacements are small enough so that the Lyapunov function is reduced at every step.

The same rationale applies to maxima, noticing that around a maximum the Hessian is negative definite and thus $\nabla V(\rho_i)^T(\rho_{i+1} - \rho_i) > 0$. □

So, unfortunately, attractiveness is not restricted to minima in the Newton-Raphson algorithm. This may sound like trouble, and to some extent it is. Because there are more attractive fixed-points, the DOA of each one tends to be smaller. As a consequence, typically the Newton-Raphson algorithm requires better initialization than the steepest descent. On the other hand, the Newton-Raphson usually presents much faster convergence and does not require any parameter commissioning, like the step sizes γ_i in the steepest descent. The following example, which consists of application of the Newton Raphson algorithm to the same objective function of Example 4.1, illustrates these facts.

Example 4.2 Consider again the following objective function

$$J(\rho) = e^{\rho}(\rho - 1)^2$$

plotted in Fig. 4.2, which presents an isolated minimum at $\rho = -1$ and an isolated maximum at $\rho = 1$. The steepest descent algorithm and the Newton-Raphson algorithm were both applied to search the minimum of the function.

The gradient of this objective function is given by

$$\nabla J(\rho) = e^{\rho}(2\rho - 2) + e^{\rho}(\rho - 1)^2$$

and its Hessian:

$$\nabla^2 J(\rho) = 2e^{\rho} + 2e^{\rho}(2\rho - 2) + e^{\rho}(\rho - 1)^2.$$

It has been seen in Example 4.1 that, provided that the step sizes are properly chosen, the global minimum is an attractive fixed-point of the steepest descent algorithm; moreover, the set $\Omega_{SD} = (-1; \infty)$ is a candidate DOA. For initial conditions outside this set the steepest descent algorithm diverges.

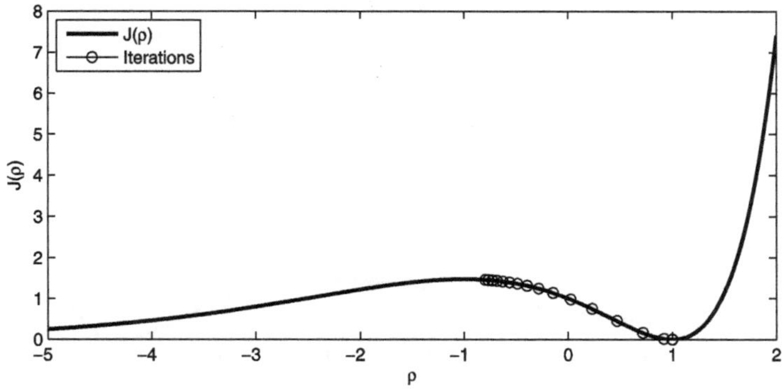

Fig. 4.7 Convergence of the steepest descent algorithm with initial condition $\rho_0 = -0.8$ and step size $\gamma_i = 0.2\ \forall i$

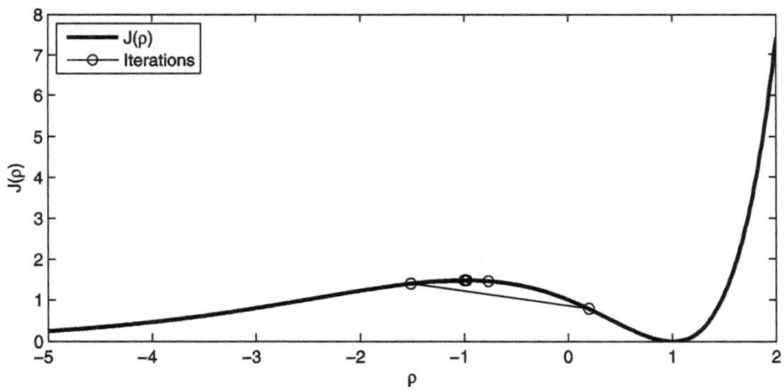

Fig. 4.8 Divergence of the Newton-Raphson method with initial condition $\rho_0 = 0.2$

With the Newton-Raphson algorithm, both the minimum and the maximum are attractive fixed-points. By exhaustive simulations it can be verified that the Newton-Raphson algorithm converges to the minimum if the initial condition is in the set $\Omega_{NR}^{\min} = (-2.41; -1.93) \cup (0.25; 0.28) \cup (0.42; \infty)$. On the other hand, the set $\Omega_{NR}^{\max}(-1.89; 0.24)$ is a DOA for the maximum. For initial conditions that do not belong to either $\Omega_{NR}^{\min}$ or $\Omega_{NR}^{\max}$ the Newton-Raphson algorithm diverges.

Some illustrative simulations are given. Figure 4.7 shows the convergence to the minimum with the steepest descent method from the initial condition $\rho_0 = -0.8$ and step size $\gamma = 0.2$. Figure 4.8 shows the convergence of the algorithm to the maximum, with the Newton-Raphson method, from the initial condition $\rho_0 = 0.2$. Figure 4.9 shows the convergence to the minimum with the Newton-Raphson method from the initial condition $\rho_0 = 0.5$. When the Newton-Raphson algorithm converges, it does so much faster (that is, within a much smaller number of iterations) than the steepest descent.

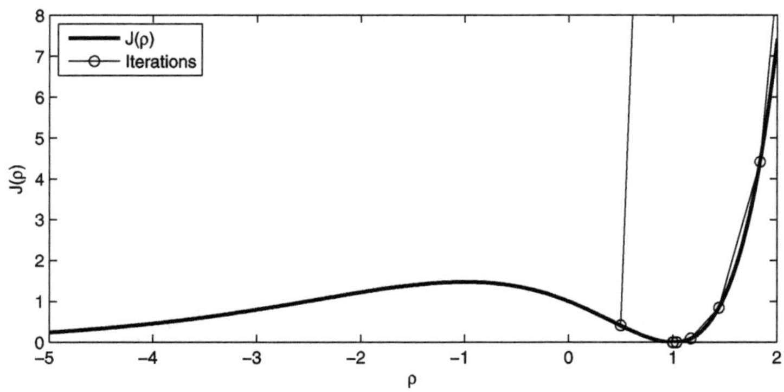

Fig. 4.9 Convergence of the Newton-Raphson method with initial condition $\rho_0 = 0.5$

All pros and cons weighted, the Newton-Raphson algorithm tends to be favored more often than the steepest descent. As an insight into the convergence of the Newton-Raphson method, consider an objective function $J(\rho)$ which is quadratic, that is:

$$J(\rho) = a + \mathbf{b}^T \rho + \rho^T C \rho$$

for some properly defined constants a, $\mathbf{b}$ and $C = C^T > 0$, and let ρ_* be *the* minimum of this function (since the function is strictly convex by definition, there is only one minimum). The gradient of this quadratic function is given by

$$\nabla J(\rho) = b + 2C\rho \tag{4.18}$$

and the minimum is the unique solution of the linear equation $\nabla J(\rho) = 0$, which is given by $\rho_* = -(2C)^{-1} b$. Adding and subtracting $2C\rho_*$ in (4.18) yields

$$\nabla J(\rho) = b + 2C\rho_* + 2C(\rho - \rho_*) = 2C(\rho - \rho_*). \tag{4.19}$$

Finally, isolating ρ_* in (4.19) yields

$$\rho_* = \rho - (2C)^{-1} \nabla J(\rho). \tag{4.20}$$

Equation (4.20) shows that, for any given ρ, the minimum ρ_* can be exactly calculated by this expression, which is exactly the expression of the NR's recursion since $\nabla^2 J(\rho) = 2C$. Hence, given any initial guess for the global minimum, the next guess, calculated by the Newton-Raphson algorithm, will be exactly the global minimum.

In conclusion, if the cost function were quadratic, then the algorithm would lead exactly to the minimum in only one iteration. Accordingly, one can expect the Newton-Raphson algorithm to converge in few iterations if the cost function is close to a quadratic function. Indeed, this is what happens, and the Newton-Raphson algorithm converges much faster than the steepest descent method in most cases, provided that the initial condition of the algorithm is sufficiently close to the minimum. However, if the initial condition is far enough from the global minimum, so that in the domain containing both the global minimum and the initial condition the cost function does not resemble a quadratic, then the algorithm looses some of its underlying logic. This is in opposition to the steepest descent, whose rationale is valid regardless of the cost function's shape. Thus, one may also expect that the steepest descent will present a larger DOA than the Newton-Raphson, and this is actually observed in practice, as will be seen in a number of examples along this book. Moreover, the Newton-Raphson method requires knowledge not only of the cost function's gradient, but also of its Hessian, whose determination can be very costly in many applications.

4.3.4 Robustness

The fundamental data necessary for the optimization algorithms presented previously are the gradient and the Hessian of the objective function. In real applications, the objective function is at best known only approximately, so these data can only be computed approximately. In some of these applications, and data-driven control design is counted among these, the objective function is not explicitly known at all, so estimation of the gradient and the Hessian is obtained indirectly, by somewhat exotic means sometimes, as will be discussed in Chap. 7. Then a question naturally arises: are the convergence properties of these algorithms robust with respect to these estimation errors? In other words, if only approximate values of the gradient (and the Hessian) are available, do the algorithms keep these properties? The answer will fortunately be yes.

Assume that the algorithm is perturbed so that

$$\rho_{i+1} = \rho_i - \gamma_i R_i \nabla J(\rho_i) + g_i \tag{4.21}$$

where g_i represents the error in the implementation of the algorithm. In the sequence we will present two robustness results with different assumptions on the error. The first one assumes that g_i is stochastic with limited variance and the second one assumes that g_i has limited amplitude.

Theorem 4.5 (Stochastic approximation) *Assume that*:

1.
$$(\rho - \rho_*)^T R_i \nabla J(\rho) > 0 \quad \forall \rho \in \mathbb{R}^p, \ \forall i$$

2.
$$E[g_i] = \mathbf{0} \quad \forall i$$

3.
$$E[(R_i \nabla J(\rho_i))^T (R_i \nabla J(\rho_i))] + E[g_i^T g_i] \le C < \infty \quad \forall \rho \in \mathbb{R}^p, \ \forall i$$

4.

$$\sum_{i=1}^{\infty} \gamma_i = \infty \qquad \sum_{i=1}^{\infty} \gamma_i^2 < \infty \tag{4.22}$$

then the algorithm (4.21) *converges* almost surely *to* ρ_*.

This theorem comes from the *Stochastic Approximation* theory, which has the paper [5] as a seminal reference. The proof of the theorem can be found in [2].

Let us analyze the assumptions of the theorem. Assumption 1 states that the angle between the algorithm direction and the vector $(\rho - \rho_*)$ should be less than $\pi/2$ rad. We have seen that this condition must be satisfied even if the algorithm is not disturbed to ensure convergence, so this is like saying that the unperturbed algorithm must converge. The nature of the disturbance is specified in Assumptions 2 and 3. The disturbance must have zero average and finite variance. Assumption 3 also implies that $\|R_i \nabla J(\rho_i)\|$ must also be finite. The final assumption specifies constraints that the step sizes must respect.

One possible choice of step sizes that respect these constraints is the harmonic sequence $\gamma_i = K/i$ with $K > 0$. This sequence of step sizes is often used in data-driven design, as well as in other optimization applications. This step size sequence ensures the convergence to a minimum, but the convergence rate is usually very low.

The following is an alternative, more recent result, with a different setting.

Theorem 4.6 [3] *Consider a discrete time system*

$$\rho_{i+1} = f(\rho_i)$$

with equilibrium point ρ_* *admitting a quadratic Lyapunov function*

$$V(\rho_i) = (\rho_i - \rho_*)^T (\rho_i - \rho_*)$$

and that $\exists \beta < 1, \alpha > 0$ *such that*

$$V(f(\rho)) < \beta^2 V(\rho), \quad \forall \rho \in \mathcal{B}_\alpha(\rho_*),$$

where

$$\mathcal{B}_\alpha(\rho_*) = \{\rho : \|(\rho - \rho_*)\| < \alpha\}.$$

Consider also that the system is perturbed by the disturbance g_i

$$\rho_{i+1} = f(\rho_i) + g_i \tag{4.23}$$

where $\|g_i\| < \delta \ \forall i$. *Consider the set*

$$\mathcal{S}(\rho_*) = \left\{\rho_i : \|(\rho - \rho_*)\| < \frac{\delta}{1-\beta}\right\}$$

and that $\mathcal{S}(\rho_*) \subset \mathcal{B}_\alpha(\rho_*)$. *Then for all* $\rho_0 \in \mathcal{B}_\alpha(\rho_*)$ *the disturbed system* (4.23) *converges to the invariant set* $\mathcal{S}(\rho_*)$.

Proof To study the convergence of the system, the quadratic Lyapunov function candidate $V(\rho)$ is used. For all $\rho_0 \in \mathcal{B}_\alpha(\rho_*)$ the disturbed system (4.23) converges to $\mathcal{S}(\rho_*)$ if

$$V(\rho_{i+1}) - V(\rho_i) < 0; \quad \forall \rho_i \in \{\mathcal{B}_\alpha(\rho_*) - \mathcal{S}(\rho_*)\}. \tag{4.24}$$

Using the conditions of the theorem $\forall \rho_i \in \mathcal{B}_\alpha(\rho_*)$

$$\begin{aligned} V(\rho_{i+1}) - V(\rho_i) &= \|(f(\rho_i) - \rho_*)\|^2 - \|(\rho_i - \rho_*)\|^2 + 2g_i^T \|(f(\rho_i) - \rho_*)\| + g_i^T g_i \\ &< \beta^2 \|(\rho_i - \rho_*)\|^2 - \|(\rho_i - \rho_*)\|^2 + 2\delta\beta \|(\rho_i - \rho_*)\| + \delta^2. \end{aligned}$$

Then a sufficient condition to ensure (4.24) is

$$(\beta \|(\rho_i - \rho_*)\| + \delta)^2 - \|(\rho_i - \rho_*)\|^2 < 0 \quad \forall \rho_i \in \{\mathcal{B}_\alpha(\rho_*) - \mathcal{S}(\rho_*)\}.$$

From the definition of $\mathcal{S}(\rho_*)$ it is easy to see that the above condition is always verified. It is still necessary to show that $\mathcal{S}(\rho_*)$ is an invariant set. From the assumptions of the theorem, if $\rho_i \in \mathcal{S}(\rho_*)$ then

$$\begin{aligned} \|\rho_{i+1} - \rho_*\| &= \|f(\rho_i) - \rho_* + g_i\| < \|f(\rho_i) - \rho_*\| + \|g_i\| \\ &< \beta\|\rho_i - \rho_*\| + \|g_i\| < \frac{\beta\delta}{1-\beta} + \delta = \frac{\delta}{1-\beta} \end{aligned}$$

which verifies that $\mathcal{S}(\rho_*)$ is invariant. □

This theorem shows that if the undisturbed system has a minimum convergence rate specified by β then the system disturbed by g_i converges to a ball of radius $\frac{\delta}{1-\beta}$ centered at the fixed-point ρ_*. If the system's state is initialized outside the ball, then it will enter the ball and never leave it again. The size of this ball is proportional to the disturbance and it is also a function of the convergence rate β. Hence, whenever an undisturbed system converges to a fixed-point, its disturbed version converges to a neighborhood of this same fixed-point.

4.4 Chapter Conclusions

The steepest descent and Newton-Raphson algorithms are probably the most basic and essential options in optimization in general; these are also the algorithms most commonly applied in data-driven control design. A wide variety of other optimization algorithms are derived from, or inspired by, the Newton-Raphson and steepest descent algorithms: Gauss-Newton, Levenberg-Marquardt, the so-called quasi-Newton family, and so forth. The fundamental properties of the Newton-Raphson and of the steepest descent have been presented in this chapter. Other algorithms will inherit these properties to the same extent that they resemble either one of these two essential algorithms—which in most cases is to a large extent.

The basic properties presented in this chapter are well-known, though they are not usually expressed in the language of Lyapunov stability as presented here. This particular approach to the problem, which has been taken from recent publications [1, 3, 4], will prove instrumental in the remaining of this book. Particularly important will be Theorem 4.3 and its constructive proof.

The two robustness theorems presented at the end of the chapter imply that the perturbed system converges to a neighborhood of the minimum provided that the disturbance satisfies some boundedness conditions and that the undisturbed system converges to the minimum. This justifies our focus on the undisturbed algorithms from now on.

In the next chapter we will particularize the convergence analysis to the objective functions that are found in data-driven control design—the ones presented in Chap. 2.

References

1. A.S. Bazanella, M. Gevers, L. Mišković, B.D.O. Anderson, Iterative minimization of H_2 control performance criteria. Automatica **44**(10), 2549–2559 (2008)
2. J.R. Blum, Multidimensional stochastic approximation methods. Ann. Math. Stat. **25**(4), 737–744 (1954)
3. D. Eckhard, A.S. Bazanella, Robust convergence of the steepest descent method for data-based control. Int. J. Syst. Sci. (2011 in press). http://www.informaworld.com/10.1080/00207721.2011.563874
4. D. Eckhard, A.S. Bazanella, Data-based controller tuning: Improving the convergence rate, in *Decision and Control (CDC), 2010 49th IEEE Conference on*, (2010), pp. 4801–4806
5. H. Robbins, S. Monro, A stochastic approximation method. Ann. Math. Stat. **22**(3), 400–407 (1951)
6. M. Vidyasagar, *Nonlinear Systems Analysis*, 2nd edn. (Prentice Hall, New York, 1992)

Chapter 5
Convergence to the Globally Optimal Controller

In the H_2 control design problem, we are looking for the best controller parameter, that is, the global minimum of the performance criterion. With that in mind, in this chapter we will apply the analytical results presented earlier to the analysis of the objective functions that we are interested in: the H_2 control performance criteria. We will explore the structure, properties and particularities of these objective functions, and we will be concerned primarily with the convergence to their global minima. This analysis will allow us to write convergence conditions that are more specific and sometimes stronger, and also to express them in ways that are directly verifiable and/or computable so that they can be used in practical data-driven control design.

We actually separate this problem into two: an analysis problem and a synthesis problem. The first one is: given a particular process, reference and performance criterion, which define uniquely the H_2 cost function, and assuming that the initial condition of the algorithm is within a candidate DOA of the global minimum, how to guarantee convergence to the global minimum and do so as fast as possible. This issue is the subject of the present chapter. It involves mainly the choice of the optimization algorithm and its parameters—the search direction and the step sizes.

With the results of this chapter under our belt we will be equipped to approach the second problem, which is the following: if the initial controller parameter is *not* inside a candidate DOA of the global optimum, what can be done about it? This question is answered by what we call *cost function shaping*, a collection of data manipulation and experimental tricks that change the cost function so that candidate DOAs containing the initial controller parameters are created. This is the subject of the following chapter.

It is never too much to stress that, up to this point and for the next two chapters, the results are not related to the particular method used to calculate the gradient and/or the Hessian used in the optimization algorithms. All these results are inherent properties of the problem formulation as the iterative optimization of an H_2 performance objective. Different methods for obtaining appropriate estimates of the gradient and the Hessian are the subject of another chapter of this book.

A. Sanfelice Bazanella et al., *Data-Driven Controller Design*,
Communications and Control Engineering,
DOI 10.1007/978-94-007-2300-9_5, © Springer Science+Business Media B.V. 2012

5.1 J_y—The Reference Tracker

In this section we will scrutinize the objective function $J_y(\rho)$, calculating convenient expressions for its gradient and then analyzing its relevant properties from the point of view of optimization. The frequency domain expression of the objective function is reproduced here for ease of reference:

$$J_y(\rho) = \frac{1}{2\pi}\int_{-\pi}^{\pi} |T(e^{j\omega},\rho) - T_d(e^{j\omega})|^2 \Phi_r(e^{j\omega})d\omega. \tag{5.1}$$

Taking the derivative of $J_y(\rho)$ in (5.1):

$$\begin{aligned}\nabla J_y(\rho) = \frac{1}{2\pi}\int_{-\pi}^{\pi} \Phi_r(e^{j\omega}) \Big\{ &[T^*(e^{j\omega},\rho) - T_d^*(e^{j\omega})]\frac{\partial}{\partial\rho}[T(e^{j\omega},\rho) - T_d(e^{j\omega})] \\ &+ [T(e^{j\omega},\rho) - T_d(e^{j\omega})]\frac{\partial}{\partial\rho}[T^*(e^{j\omega},\rho) - T_d^*(e^{j\omega})]\Big\} d\omega \end{aligned} \tag{5.2}$$

where the superscript * indicates the complex conjugate of a complex number.

This expression is not really convenient for analysis, because of the rational dependence of $T(e^{j\omega},\rho)$ on the unknown parameter ρ. So, let us develop it into something more useful, starting with

$$\begin{aligned}T(e^{j\omega},\rho) - T_d(e^{j\omega}) &= \frac{C(e^{j\omega},\rho)G(e^{j\omega})}{1+C(e^{j\omega},\rho)G(e^{j\omega})} - \frac{C_d(e^{j\omega})G(e^{j\omega})}{1+C_d(e^{j\omega})G(e^{j\omega})} \\ &= (C(e^{j\omega},\rho) - C_d(e^{j\omega}))G(e^{j\omega})S_d(e^{j\omega})S(e^{j\omega},\rho)\end{aligned} \tag{5.3}$$

where we have defined the "desired sensitivity function" $S_d(z,\rho) \triangleq 1 - T_d(z,\rho)$, which is the sensitivity function when the closed loop behaves exactly as desired. Similarly:

$$(T(e^{j\omega},\rho) - T_d(e^{j\omega}))^* = (C(e^{j\omega},\rho) - C_d(e^{j\omega}))^* G^*(e^{j\omega})S_d^*(e^{j\omega})S^*(e^{j\omega},\rho). \tag{5.4}$$

On the other hand,

$$\frac{\partial T(e^{j\omega},\rho)}{\partial\rho} = \frac{\partial}{\partial\rho}\frac{C(e^{j\omega},\rho)G(e^{j\omega})}{1+C(e^{j\omega},\rho)G(e^{j\omega})} = G(e^{j\omega})S^2(e^{j\omega},\rho)\frac{\partial C(e^{j\omega},\rho)}{\partial\rho} \tag{5.5}$$

and similarly:

$$\frac{\partial T^*(e^{j\omega},\rho)}{\partial\rho} = G^*(e^{j\omega})S^{*2}(e^{j\omega},\rho)\frac{\partial C^*(e^{j\omega},\rho)}{\partial\rho}. \tag{5.6}$$

Inserting (5.3), (5.4), (5.5) and (5.6) in (5.2) yields, after some simplification:

$$\begin{aligned}\nabla J_y(\rho) = \frac{1}{\pi}\int_{-\pi}^{\pi} &\Phi_r(e^{j\omega})|G(e^{j\omega})S(e^{j\omega},\rho)|^2 \\ &\times \Re\left\{(C(e^{j\omega},\rho) - C_d(e^{j\omega}))^* S_d^*(e^{j\omega})S(e^{j\omega},\rho)\frac{\partial C(e^{j\omega},\rho)}{\partial\rho}\right\} d\omega\end{aligned} \tag{5.7}$$

where $\Re\{\cdot\}$ indicates the real part of a complex quantity.

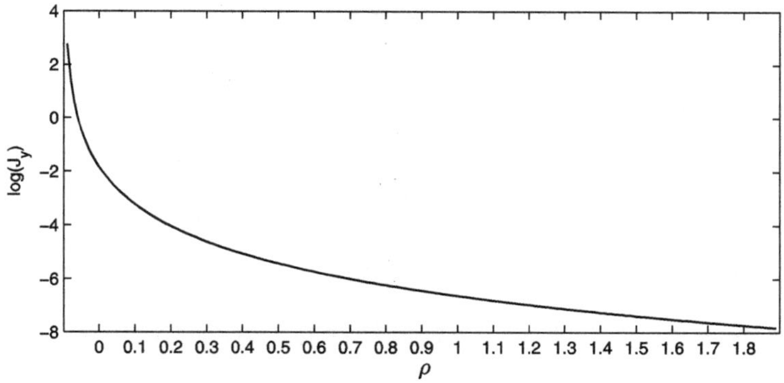

Fig. 5.1 Objective function $J_y(\rho)$ for Example 5.1. The function's value declines up to $\rho = 1.9$, which is at the border of the stability set $\Gamma = (-0, 1; 1.9)$

Notice in this expression that if $C(z, \rho) = C_d(z)$ then the gradient is zero. This comes as no surprise, of course, for the value of ρ for which $C(z, \rho) = C_d(z)$, if such a ρ exists, is a global minimum. But also notice that if Assumption B_y is not satisfied, that is, there is no value of ρ for which $C(z, \rho) = C_d(z)$, then the global minimum depends on the spectrum of the reference $\Phi_r(e^{J\omega})$. It may even happen that the gradient is never zero.

Example 5.1 Consider the very simple example of a first order process controlled by a proportional controller which must track a constant reference

$$G_0(z) = \frac{1}{z - 0.9} \qquad C(z, \rho) = \rho \qquad T_d(z) = 1 \qquad r(t) \equiv 1$$

with $\rho \in \mathbb{R}^+$. The closed-loop system behaves as

$$T(z, \rho) = \frac{\rho}{z - 0.9 + \rho}$$

and we observe that Assumption B_y is not satisfied. We also observe that

$$\lim_{\rho \to \infty} T(z, \rho) = T_d(z)$$

so that one must be tempted to make ρ as large as possible to approach the desired closed-loop performance $T_d(z)$. But the system becomes unstable for $\rho > 1.9$, which implies that for such values of the parameter ρ the value of the objective function becomes infinity. So we can expect that the objective function will decrease as $\rho \to 1.9$, and then present a discontinuity at this value. Indeed, this is what happens, as can be seen in Fig. 5.1, which presents the objective function $J_y(\rho)$ evaluated at different values of ρ.

This model reference design is not a well posed problem, because it does not have a solution. Indeed, there is not a value of ρ that satisfies the definition of

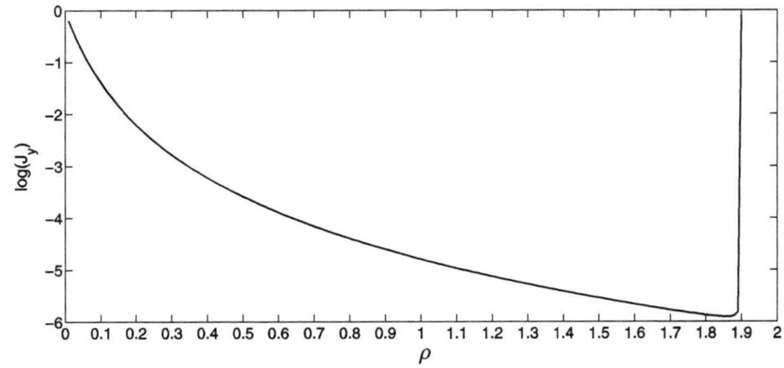

Fig. 5.2 Approximation of the cost function $J_y(\rho)$ for Example 5.1 obtained with $N = 10^5$ data

a minimum: for any value $\rho' < 1.9$ there always exists a $\rho'' \in (\rho'; 1.9)$ such that $J_y(\rho'') < J_y(\rho')$.

In practice the situation is a bit different, because the objective function $J_y(\rho)$ is never computed or optimized exactly; the function that is actually computed and thus minimized is an approximation computed with a finite set of data (for details see Sect. 6.4.3 and Chap. 7, where computations are discussed). This function does possess a global minimum in Γ, as can be seen in Fig. 5.2, in which the function obtained for $N = 10^5$ data is plotted, but this does not make the situation any better from a practical point of view.

Strictly speaking, the optimization of the approximate function is well posed, because this function possesses a global minimum. But this minimum can be arbitrarily close to the stability boundary, so that from a practical point of view the situation is as bad as if it did not exist. Any iterative optimization algorithm looking for a solution that is arbitrarily close to the stability boundary is likely to skip it and end up at the other side of the boundary at some iteration, thus resulting in an unstable closed loop. And even assuming that some algorithm can find this global minimum without ever trespassing the stability limits, it would be by definition a bad solution. So, whether talking about the asymptotic cost function $J_y(\rho)$ or its finite data approximation, the bottom line for this example is the same: the control objective specified is not compatible with the controller structure available, causing an inconsistency in the H_2 problem.

At this point we hope to have given the reader some motivation to choose the reference model such that Assumption B_y is satisfied. But more will come along the book, as we establish a variety of nice properties of the H_2 control design problem that are present only when this assumption is verified or mildly violated. Let us also remind the reader that this assumption is equivalent to knowing a model class to which the process model belongs, as discussed in Chap. 2 (see Examples 2.1 and 2.4).

Let us move on from (5.7). From now on we treat the case of a linearly parameterized controller, in the terms of Assumption LP in Chap. 1, which says that the controller's transfer function can be written like:

$$C(z,\rho) = \rho^T \bar{C}(z) \tag{5.8}$$

As discussed before, this covers most practical cases and even when the actual controller is not in the form (5.8), an appropriate basis $\bar{C}(z)$ can be chosen to approximate the controller's transfer function to any desired precision. The linear parameterization form (5.8) implies that

$$\frac{\partial C(z,\rho)}{\partial \rho} = \bar{C}(z)$$

$$\frac{\partial C^*(z,\rho)}{\partial \rho} = \bar{C}^{*T}(z)$$

where * has the usual meaning of complex conjugate transpose, that is, $\bar{C}^*(e^{J\omega}) = \bar{C}^T(e^{-j\omega})$. Under this assumption,

$$\begin{aligned}\nabla J_y(\rho) = \frac{1}{\pi}\int_{-\pi}^{\pi} &\Phi_r(e^{J\omega})|G(e^{J\omega})S(e^{J\omega},\rho)|^2 \\ &\times \Re\{(C(e^{J\omega},\rho) - C_d(e^{J\omega}))^* S_d(e^{J\omega})^* S(e^{J\omega},\rho)\bar{C}(e^{J\omega})\} d\omega.\end{aligned} \tag{5.9}$$

If Assumption B_y is also satisfied, then we can write

$$C(z,\rho) - C_d(z) = (\rho - \rho_d)^T \bar{C}(z). \tag{5.10}$$

which inserted into (5.9) results in

$$\begin{aligned}\nabla J_y(\rho) = \frac{1}{\pi}\int_{-\pi}^{\pi} &\Phi_r(e^{J\omega})|G(e^{J\omega})S(e^{J\omega},\rho)|^2 \\ &\times \Re\{S_d^*(e^{J\omega})S(e^{J\omega},\rho)\bar{C}(e^{J\omega})\bar{C}^*(e^{J\omega})\} d\omega(\rho - \rho_d).\end{aligned} \tag{5.11}$$

Finally, defining the matrix

$$\begin{aligned}M(\rho) \triangleq \frac{1}{\pi}\int_{-\pi}^{\pi} &\Phi_r(e^{J\omega})|G(e^{J\omega})|^2|S(e^{J\omega},\rho)|^2 \\ &\times \Re\{S_d^*(e^{J\omega})S(e^{J\omega},\rho)\bar{C}(e^{J\omega})\bar{C}^*(e^{J\omega})\} d\omega\end{aligned} \tag{5.12}$$

(5.11) can be written as

$$\nabla J_y(\rho) = M(\rho)(\rho - \rho_d). \tag{5.13}$$

As a result of the linear parameterization and of the matched control assumptions, the gradient of the cost function can be written in the form (5.13). This is a convenient expression to study the convergence of the optimization methods because then

$$(\rho - \rho_d)^T \nabla J(\rho) = (\rho - \rho_d)^T M(\rho)(\rho - \rho_d), \tag{5.14}$$

which is a quadratic form.

Convergence of an optimization algorithm from an initialization in a given set can be ensured by the satisfaction of condition (4.12) within this set, which for the particular cost function $J_y(\rho)$ assumes the quadratic form (5.14). And whether this condition is satisfied in a given set depends on the properties of the matrix $M(\rho)$ for all ρ contained in this set; let us examine these properties. From (5.12):

$$\begin{aligned} M(\rho) = & \frac{1}{\pi}\int_{-\pi}^{\pi} \Phi_r(e^{j\omega})|G(e^{j\omega})|^2|S(e^{j\omega},\rho)|^2 \Re\{S_d^*(e^{j\omega})S(e^{j\omega},\rho)\} \\ & \times \Re\{\bar{C}(e^{j\omega})\bar{C}^*(e^{j\omega})\}d\omega \\ & - \frac{1}{\pi}\int_{-\pi}^{\pi} \Phi_r(e^{j\omega})|G(e^{j\omega})|^2|S(e^{j\omega},\rho)|^2 \\ & \times \Im\{S_d^*(e^{j\omega})S(e^{j\omega},\rho)\}\Im\{\bar{C}(e^{j\omega})\bar{C}^*(e^{j\omega})\}d\omega \\ \triangleq & M_s(\rho) + M_a(\rho). \end{aligned} \tag{5.15}$$

It is straightforward to verify that $M_a(\rho)$ is anti-symmetric, whereas $M_s(\rho)$ is symmetric—hence their subscripts. Since $x^T Q x = 0$ for any anti-symmetric matrix Q and any $x \in \mathbb{R}^n$, (4.14) can be written as

$$(\rho - \rho_d)^T M_s(\rho)(\rho - \rho_d) > a\|\rho - \rho_d\| \|\nabla J_y(\rho)\| \quad \forall \rho \in \mathcal{B}_\alpha(\rho_d) \setminus \rho_d. \tag{5.16}$$

Assume (the standard assumption) that the cost function $J_y(\rho)$ is Lipschitz continuous around the global minimum, that is,

$$\|\nabla J_y(\rho)\| < b\|\rho - \rho_d\| \quad \forall \rho \in \mathcal{B}_\alpha(\rho_d) \setminus \rho_d$$

for some positive $b \in \mathbb{R}^+$. Then condition (5.16) is satisfied if

$$M_s(\rho) > \varepsilon > 0$$

where $\varepsilon = ab$. That is, a ball $\mathcal{B}_\alpha(\rho_d)$ is a candidate DOA if, for some $\varepsilon > 0$, the symmetric matrix $M_s(\rho) - \varepsilon I$ is positive definite for all parameter values ρ in this ball.

In order to simplify the presentation, we would rather work with the simpler condition $M_s(\rho) > 0$ in lieu of $M_s(\rho) - \varepsilon I > 0$ for some ε. Thanks to the continuity of all the functions involved, the technical difference between these two conditions can be taken care of by always verifying and enforcing the positivity of $M_s(\rho)$ condition in a closed set. Indeed, the existence of $\varepsilon > 0$ such that $M_s(\rho) - \varepsilon I > 0$ $\forall \rho \in \Omega$ in an open set Ω is equivalent to $M_s(\rho) > 0$ $\forall \rho \in \bar{\Omega}$.[1] Then we have the following result.

Lemma 5.1 *A ball $\mathcal{B}_\alpha(\rho_d)$ is a candidate DOA if $M_s(\rho) > 0$ $\forall \rho \in \bar{\mathcal{B}}_\alpha(\rho_d) \setminus \rho_d$.*

[1] $\bar{\Omega}$ being the closure of Ω.

So, from now on we will focus on the condition $M_s(\rho) > 0$. This positive definiteness condition will be dissected by examining the expression (5.15). At any given frequency ω_i in the support of $r(t)$, the scalar factor in the integrand $\Phi_r(e^{\jmath\omega_i})|G(e^{\jmath\omega_i})|^2|S(e^{\jmath\omega_i},\rho)|^2$ is positive, except for frequencies ω_i corresponding to zeros on the unit circle of the process or of the sensitivity, that may or may not exist. On the other hand, the matrix $\Re\{\bar{C}(e^{\jmath\omega_i})\bar{C}^*(e^{\jmath\omega_i})\}$ is a positive semi-definite matrix. Hence the only factor in the integrand that can be negative is $\Re\{S_d^*(e^{\jmath\omega_i})S(e^{\jmath\omega_i},\rho)\}$. Assuming that $\Re\{S_d^*(e^{\jmath\omega})S(e^{\jmath\omega},\rho)\}$ is positive for all ω, each term of the sum forming the matrix $M_s(\rho)$ will be positive semi-definite. Then, as discussed in Chap. 3, the sum of p (the dimension of the parameter vector ρ) such positive semi-definite matrices will result in a full-rank (thus positive definite) matrix $M_s(\rho)$ as a generic property, that is, for almost all reference signals containing at least p nonzero frequencies is its support. With this argument we have proven the following theorem.

Theorem 5.1 *Let Assumptions* B_y, LP *and* LI *be satisfied and let* $\Upsilon \subseteq \Gamma$ *be a connected set such that* $\rho_d \in \Upsilon$ *and, for all* $\rho \in \Upsilon$:

$$\Re\{S_d^*(e^{\jmath\omega})S(e^{\jmath\omega},\rho)\} > 0 \quad \forall\omega. \tag{5.17}$$

Then, for almost all SRp references $r(t)$:

$$(\rho - \rho_d)^T \nabla J_y(\rho) > 0 \quad \forall \rho \in \Upsilon \setminus \rho_d.$$

This theorem is the cornerstone of the convergence results that follow, and also of the methodologies of cost function shaping, which will be presented in a subsequent chapter. We have seen in Chap. 4 that with the satisfaction of condition (4.12) in a given set, the optimization "tends to be easy" from initial conditions within this set. By "easy" we mean that the steepest descent converges to the global optimum, and there is no need for sophisticated optimization algorithms or ad-hoc schemes. With Theorem 5.1 that generic condition, which is obviously not computable because it depends on the global optimum ρ_d, has been translated into the Strictly Positive Real (SPR) condition (5.17), which is related directly to the desired closed loop performance and the set that is a candidate to be a DOA. A number of specific MRAC methods, as well as some data-driven methods, arrive at similar SPR conditions for convergence, although usually through very different paths.

Interpretation of the conditions of this theorem is a major step towards the appropriate use of data-driven and adaptive control methods; let us interpret them one by one. First note that these conditions guarantee that $M_s(\rho)$ is positive definite, which is sufficient for the positivity condition (4.12) to be satisfied; it is, in principle, not necessary. Indeed, even if at a particular value of ρ the matrix is not positive definite, the positivity condition (4.12) can still be positive, depending on the direction of vector $\rho - \rho_d$. But the direction of this vector is not known, because it is determined by the global optimum ρ_d. So, it seems reasonable to try to guarantee the positivity condition (4.12) regardless of what ρ_d might be, which can be guaranteed only if the matrix $M_s(\rho)$ is positive definite everywhere in the set Υ.

The richness condition on the reference signal and the linear independence condition on the controller parameterization are necessary conditions that have been explained earlier. In setting up his/her data-driven control design, the designer must make sure that these conditions are satisfied. The controller parameterization is a designer's choice, and choosing an independent parameterization is the natural thing to do anyways, so in most cases this is a nonissue. The reference richness, on the other hand, may be an issue, and it may be necessary to add some extra excitation on top of the desired reference during parameter adaptation procedures in order to guarantee its richness. The "for almost all" part of the theorem statement also deserves some attention: in counting the number of frequencies of the reference in order to check the appropriate signal richness one must exclude those frequencies that are zeros of the process and those that are zeros of the sensitivity. While the first ones very seldom exist, the zeros of the sensitivity usually are exactly at the dominant frequencies of the reference—the frequencies to be tracked with zero steady-state error. So the actual reference usually must have p frequencies *besides* the ones to be tracked with zero steady-state error. These conditions and the considerations above are typical requirements in the similar problem of system identification, and for similar reasons.

Now let us take a closer look at condition (5.17). This condition can be expressed in alternative forms with some basic manipulation using elementary properties of complex numbers. First, remember that the real part of a complex number is positive if and only if its argument is in the range $(-\pi/2;\ \pi/2)$, so

$$\Re\{S_d^*(e^{j\omega})S(e^{j\omega},\rho)\} > 0 \leftrightarrow \angle\{S_d^*(e^{j\omega})S(e^{j\omega},\rho)\} \in (-\pi/2;\ \pi/2).$$

Moreover, for any complex numbers a and b, $\angle a.b = \angle a + \angle b$ and $\angle a^* = -\angle a$, so that

$$\Re\{S_d^*(e^{j\omega})S(e^{j\omega},\rho)\} > 0 \leftrightarrow \angle S(e^{j\omega},\rho) - \angle S_d(e^{j\omega}) \in (-\pi/2;\ \pi/2).$$

As a result, condition (5.17) can be expressed in the equivalent form:

$$\max_{\omega} |\angle S(e^{j\omega},\rho) - \angle S_d(e^{j\omega})| < \pi/2 \quad \forall \rho \in \Upsilon. \tag{5.18}$$

This *maximum phase difference* between the two transfer functions is a metric that can be used to measure the distance between a given ρ and the global optimum ρ_d. With this in mind, condition (5.17) reads intuitively as an obvious fact: the optimization can converge if it is initialized close enough from the global optimum. What is very far from obvious, and is told by Theorem 5.1, is how to measure this "closeness" and just how close is "close enough". Indeed, this theorem provides the appropriate metric to measure the distance, which is

$$\text{dist}(\rho^1,\rho^2) \triangleq \max_{\omega} |\angle S(e^{j\omega},\rho^1) - \angle S(e^{j\omega},\rho^2)|.$$

It also tells what is the largest distance acceptable between the initial controller and the optimal controller in order to obtain convergence.

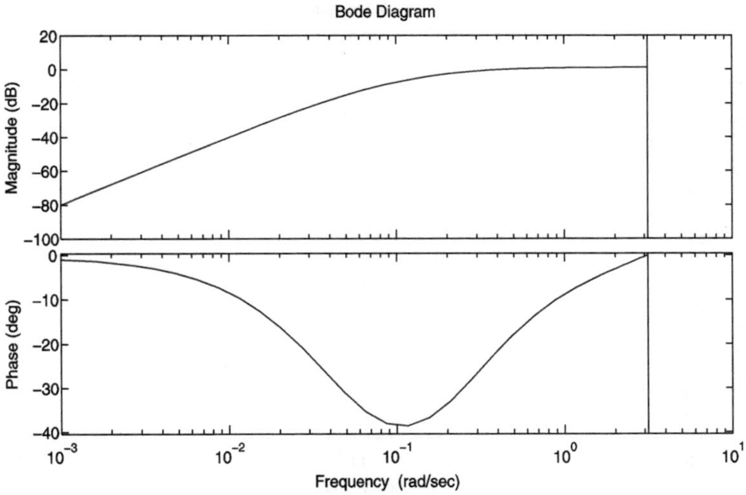

Fig. 5.3 Frequency response of $\frac{S_d(e^{J\omega})}{S(e^{J\omega},\rho^1)}$: the maximum value of its phase represents the distance $\operatorname{dist}(\rho^1, \rho_d)$

Yet another alternative, and also equivalent condition, is that the transfer function below is strictly positive real (SPR):

$$\frac{S_d(e^{J\omega})}{S(e^{J\omega}, \rho)}. \tag{5.19}$$

Note that this transfer function $\frac{S_d(e^{J\omega})}{S(e^{J\omega},\rho)}$ is always proper but never strictly proper; it is affine in ρ, and BIBO-stable for all $\rho \in \Gamma$. This alternative form will be instrumental in the computations in some of the examples to follow.

Example 5.2 Let

$$G(z) = \frac{1}{z-0.9}, \qquad C(z,\rho) = [\varrho_1\ \varrho_2]\begin{bmatrix} \frac{z}{z-1} \\ \frac{1}{z-1} \end{bmatrix}, \qquad T_d(z) = \frac{0.2}{z-0.8}. \tag{5.20}$$

The ideal controller, which minimizes $J_y(\rho)$, is achieved for $\rho_d = [0.2\ {-0.18}]^T$. Let us compare $\rho^1 = [0.05\ {-0.045}]^T$ and $\rho^2 = [0.3\ {-0.9}]^T$; which one is a better initial condition for optimization? The Euclidean distance from the initial condition ρ^1 to the global optimum ρ_d is: $\|\rho^1 - \rho_d\| = 0.2018$, whereas $\|\rho^2 - \rho_d\| = 0.1345$, so one might be tempted to say that initializing the optimization at ρ^2 is a better choice than initializing it at ρ^1. But Figs. 5.3 and 5.4 show the frequency responses of the transfer functions $\frac{S_d(e^{J\omega})}{S(e^{J\omega},\rho^1)}$ and $\frac{S_d(e^{J\omega})}{S(e^{J\omega},\rho^2)}$. The maximum phase of each transfer function represents the distance between the corresponding value of ρ and the global optimum ρ_d, so it is verified in these figures that $\operatorname{dist}(\rho^1, \rho_d) = 38.7°$ whereas $\operatorname{dist}(\rho^2, \rho_d) = 100.8°$. So ρ^1 is actually much closer to the global optimum

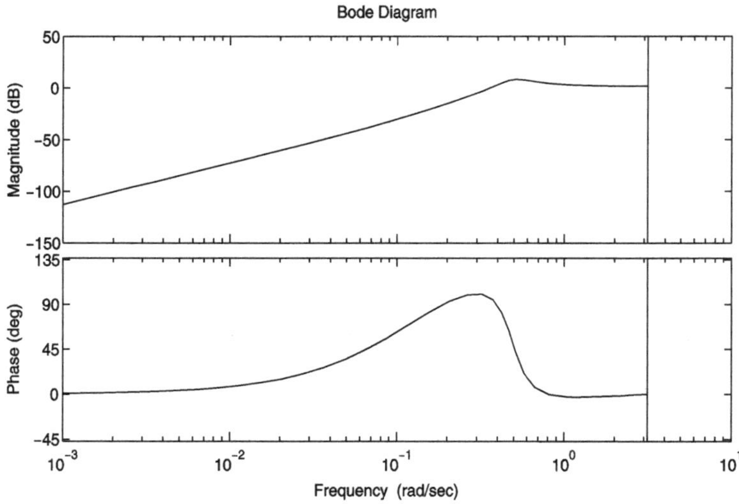

Fig. 5.4 Frequency response of $\frac{S_d(e^{j\omega})}{S(e^{j\omega},\rho^2)}$: the maximum value of its phase represents the distance $\text{dist}(\rho^2, \rho_d)$

in the appropriate sense than ρ^2 and thus it is a better initialization. Moreover, ρ^2 is not "close enough" to ρ_d, since $\text{dist}(\rho^2, \rho_d) > 90°$.

Checking the Convergence Condition Checking the SPR condition (5.18) in a given design apparently requires knowledge of the phase of the sensitivity functions $S_d(z)$ and $S(z, \rho)$, which in turn seems to demand knowledge of the process' transfer function; fortunately this is not the case. The desired sensitivity $S_d(z)$ is given as a designer's choice. As for the sensitivity $S(z, \rho)$, it is not necessary to know it, but only to have some information about its phase, an information which can present itself in different ways. One obvious possibility to assess this information is to estimate the sensitivity phase from the measurement of the signals $r(t)$ and $y(t)$ from the system, using the following relationship:

$$r(t) - y(t) = S(z, \rho)r(t).$$

This may sound like a model identification, but it is not, because only a rough estimate of the phase will be enough, and only in a frequency range where the maximum occurs. Another way to assess condition (5.18) is illustrated in the following example.

Example 5.3 Let

$$G(z) = \frac{1}{z-a}, \quad C(z, \rho) = \rho\frac{z-a}{z-1}, \quad |a| < 1.$$

The ideal controller $C_d(z)$ belongs to the controller class considered if and only if the reference model is of the form $T_d(z) = \frac{1-b}{z-b}$; then $\rho_d = 1 - b$. The sensitivity function is given by

$$S(z,\rho) = \frac{1}{1 + C(z,\rho)G(z)} = \frac{z-1}{z-(1-\rho)}.$$

Condition (5.18) will be studied through its equivalent form (5.19) involving the SPR property of the transfer function formed by the ratio of the two sensitivities. For two arbitrary sensitivity functions generated by $\rho^1, \rho^2 \in \Gamma$, we have

$$\frac{S(z,\rho^1)}{S(z,\rho^2)} = \frac{z-(1-\rho^2)}{z-(1-\rho^1)}.$$

The set of stability parameters ρ is $\Gamma = \{\rho : |1-\rho| < 1\}$. Let us define $\alpha_i = 1-\rho^i$ for convenience of notation. From the positive real lemma, this function is SPR if and only if there exists $q \in \mathbb{R}^+$ satisfying the inequalities:

$$\begin{aligned} q(1-\alpha_2^2) &> 0 \\ (\alpha_2-\alpha_1)^2 q^2 + 2(\alpha_1\alpha_2 - 1)q + 1 \triangleq \eta(q) &< 0. \end{aligned}$$

The first inequality requires $\alpha_2 < 1$, which is the case for $\rho^2 \in \Gamma$. The second one will be satisfied for some real positive q if and only if the roots of the polynomial $\eta(q)$ are real and at least one of them is positive. The roots of $\eta(q)$ are given by

$$q = \frac{-(\alpha_1\alpha_2 - 1) \pm \sqrt{(\alpha_1^2-1)(\alpha_2^2-1)}}{(\alpha_2-\alpha_1)^2}.$$

These roots are real if and only if $(\alpha_1^2-1)(\alpha_2^2-1) \geq 0$, which is satisfied for all $\rho^1, \rho^2 \in \Gamma$. Moreover, for $\rho^1, \rho^2 \in \Gamma$, $-(\alpha_1\alpha_2 - 1) > 0$, so one of the roots is positive. Hence, whatever reference model $T_d(z)$ we choose such that it is BIBO-stable and can be achieved exactly with the controller class considered, Γ will be a candidate DOA for the global minimum $\rho_d = 1-b$.

In the example, the knowledge of the model structure alone is enough to verify that Assumption B_y can be satisfied with a PI controller. In order to actually choose a controller class that satisfies Assumption B_y we need to know also the pole value a, but not the gain, that for this process equals one. This knowledge, on the other hand, allows also to characterize the whole set of reference models for which Assumption B_y is satisfied, and to verify that condition (5.17) is satisfied for all these reference models and for the whole stability set Γ—a very strong result. So, although some information on the process must be available, this required information is far less than what is required for model-based design—namely, knowledge of $G(z)$ to a reasonable degree of accuracy. Let us stress once again the obvious yet often forgotten truth that in order to guarantee anything about the control of any process it is necessary to have some information about it.

The Role of Φ_r When the necessary richness and LI conditions in Theorem 5.1 are satisfied, the condition (5.18) is sufficient for the matrix $M_s(\rho)$ to be positive definite. It is, however, far from necessary. The rationale behind this condition is to guarantee that the integral forming the matrix $M_s(\rho)$ is the sum of strictly positive terms. If this condition is not satisfied, then one or more of these terms is negative, and then it may happen that the sum is not positive definite; but it just "may".

Example 5.4 Let

$$G(z) = \frac{1}{z-0.5}, \qquad C(z,\rho) = \rho\frac{z}{z-0.9}, \qquad T_d(z) = \frac{2.4z}{z^2+z+0.45}. \tag{5.21}$$

The ideal controller, which minimizes $J_y(\rho)$, is achieved for $\rho_d = 2.4$. The stability set is $\Gamma = (-0.05,\ 2.85)$. It is straightforward to verify that $\frac{S_d(e^{j\omega})}{S(e^{j\omega},\rho)}$ is SPR for all $\rho \in \Upsilon = (1.35,\ 2.85)$. So, from Theorem 5.1, the cost $J_y(\rho)$ has no other extrema than ρ_d within the set Υ and this set is a candidate DOA for $J_y(\rho)$. For $\rho \notin \Upsilon$ the *SPR* condition is not satisfied, so there may exist local minima or maxima outside Υ. Whether such extrema exist depends on the spectrum of the reference applied to the system.

Let us reexamine (5.16) in the light of the final remark in the example above, noting that $M(\rho)$ is a scalar in that particular example. If all the factors inside the integral that forms $M_s(\rho)$ are positive, which is equivalent with the satisfaction of condition (5.17), then the integral cannot be zero and (5.16) is satisfied. For parameter values such that (5.17) is not satisfied, the matrix $M_s(\rho)$ is a sum of positive and negative terms, which may result in this matrix not being positive definite. But if the negative terms are small compared to the positive ones, then the sum will still be positive. So, if (5.17) is violated only at those frequencies where the input power (Φ_r) is low, then the integral is likely to be positive even though the SPR condition (5.17) is not satisfied. This is a powerful idea in adaptive control which will be further explored in the sequel, particularly in Chap. 6.

When condition (5.17) is not satisfied, then the factor $\Re\{S_d^*(e^{j\omega})S(e^{j\omega},\rho)\}$ is negative in a range of frequencies and positive in another range, and clearly there exists a Φ_r such that it "weighs" equally these two frequency ranges, thus causing the integral to vanish. The following corollary results immediately from this argument.

Corollary 5.1 *Let Assumptions* B$_y$, LP *and* LI *be satisfied and consider a given set* $\Upsilon \subseteq \Gamma$, *with* $\rho_d \in \Upsilon$. *If for some* ρ^1, $\exists\omega :\ \Re\{\frac{S_d(e^{j\omega})}{S(e^{j\omega},\rho^1)}\} < 0$ *then there exist SRp reference signals* $r(t)$ *such that* $\nabla J_y(\rho^1) = \mathbf{0}$. *This, in turn, implies that for such reference signals any set* $\Upsilon \ni \rho^1$ *is* not *a candidate DOA for* $J_y(\rho)$.

Example 5.5 Consider again the system of Example 5.4. The SPR condition is not satisfied for $\rho < 1.35$. Consider $\rho = 0.5$, for instance. Figures 5.5 and 5.6 illustrate that for this value of ρ the SPR condition is indeed not satisfied, since there are frequencies at which the phases of $S_d(z)$ and $S(z, 0.5)$ differ by more than $\frac{\pi}{2}$ rad.

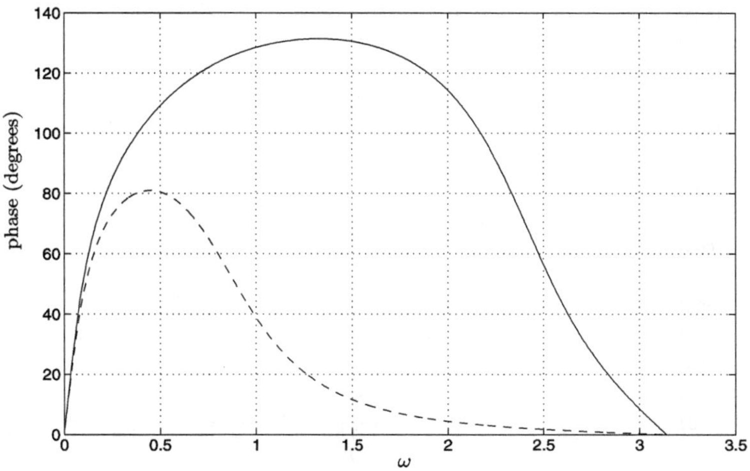

Fig. 5.5 Example 5.5—frequency response of $S_d(e^{j\omega})$ (*full line*) and $S(e^{j\omega}, \rho)$ (*dashed line*) for $\rho = 0.5$ ($\rho_d = 2.4$)

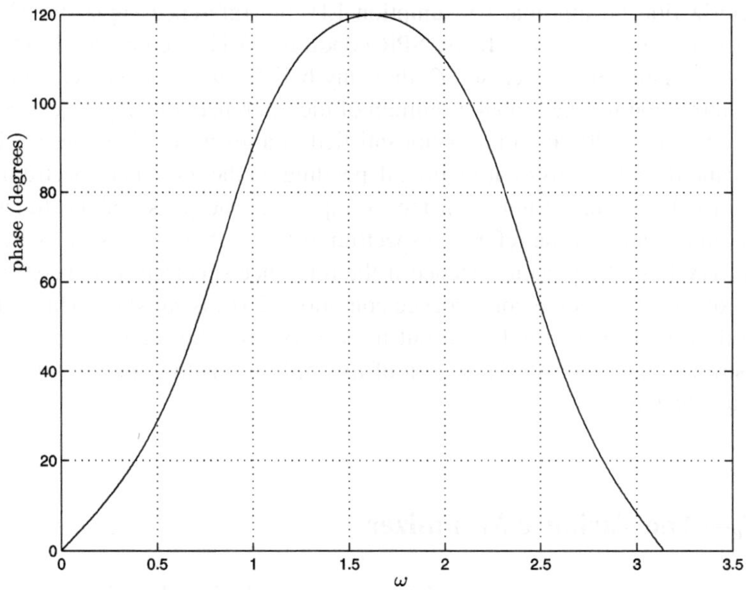

Fig. 5.6 Example 5.5—frequency response of $S_d(e^{j\omega})/S(e^{j\omega}, \rho)$

Moreover, it can be seen in these figures that $\Re\{\frac{S_d(e^{j1})}{S(e^{j1},0.5)}\} = 0$, so if only this frequency ($\omega = 1$ rad/s) is excited, the gradient will be zero at this particular value of ρ. Indeed, by applying $r(t) = \sin(1 \cdot t)$ the objective function behaves as presented in Fig. 5.7, which has a local maximum at $\rho = 0.5$.

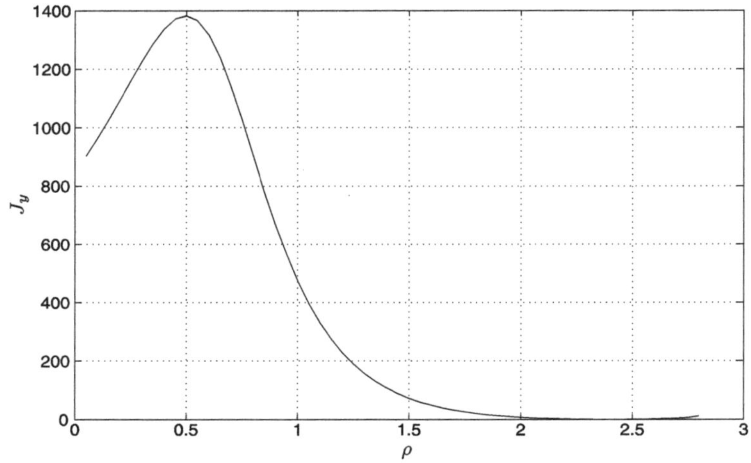

Fig. 5.7 Cost $J_y(\rho)$: global optimum at $\rho_d = 2.4$ and a maximum at $\rho = 0.5$

In short, what we have shown in this section is the following. Let the controller's parameterization be minimal (Assumption LI), the reference signal be SRp and $C_d(z) \in \mathcal{C}$ (Assumption B_y). If the SPR condition (5.17)—equivalently (5.18) or (5.19)—is satisfied in a given set Υ, then any ball $\mathcal{B}_\epsilon(\rho_d) \subset \Upsilon$ is a candidate domain of attraction for the global minimum of the reference tracking objective function $J_y(\rho)$. If the SPR condition is not satisfied in a given set, then these balls can still be candidate domains of attraction, depending on the reference spectrum.

It is also important to notice in the example that the phase of the function is small at both "ends" of the reference spectrum (see Fig. 5.6), that is, for very small and for very large frequencies. Hence if the reference spectrum is concentrated at these frequency ranges the convergence condition (5.16) is satisfied. This is in fact a general property which will turn out to be very important, but we postpone its demonstration and a deeper discussion of its implications to where it will be most useful, in Chap. 6.

5.2 J_e—The Variance Minimizer

The same treatment just given for $J_y(\rho)$ is now applied to the other performance criterion—$J_e(\rho)$. Start by calculating its gradient.

$$\nabla J_e(\rho) = \sigma_e^2 \frac{1}{2\pi} \int_{-\pi}^{\pi} |H(e^{j\omega})|^2 \frac{\partial}{\partial \rho} |S(e^{j\omega}, \rho) - S_e(e^{j\omega})|^2 d\omega \tag{5.22}$$

$$\frac{\partial}{\partial \rho} S(e^{j\omega}, \rho) = \frac{\partial}{\partial \rho} \frac{1}{1 + C(e^{j\omega}, \rho) G(e^{j\omega})} = -G(e^{j\omega}) S(e^{j\omega}, \rho)^2 \frac{\partial C(e^{j\omega}, \rho)}{\partial \rho}. \tag{5.23}$$

From (5.23) we get:

$$\begin{aligned}
&\frac{\partial}{\partial\rho}|S(e^{j\omega},\rho)-S_e(e^{j\omega})|^2\\
&\quad=-2|S(e^{j\omega},\rho)|^2|G(e^{j\omega})|^2\\
&\quad\quad\times\Re\left\{S^*(e^{j\omega},\rho)S_e(e^{j\omega})(C_e(e^{j\omega})-C(e^{j\omega},\rho))\frac{\partial C^*(\rho)}{\partial\rho}\right\}.
\end{aligned}\tag{5.24}$$

Now, assume that the controller is linearly parameterized and that the ideal controller $C_e(z)$ belongs to the controller class considered (Assumption B_e); then:

$$\begin{aligned}
C_e(z)&=\rho_e^T\bar{C}(z)\\
C(z,\rho)&=\rho^T\bar{C}(z)\\
\frac{\partial C(\rho)}{\partial\rho}&=\bar{C}(z)
\end{aligned}$$

and (5.24) becomes:

$$\begin{aligned}
\frac{\partial}{\partial\rho}|S(e^{j\omega},\rho)-S_e(e^{j\omega})|^2&=2|S(e^{j\omega},\rho)|^2|G(e^{j\omega})|^2\\
&\quad\times\Re\{S^*(e^{j\omega},\rho)S_e(e^{j\omega})\bar{C}(e^{j\omega})\bar{C}^*(e^{j\omega})\}(\rho-\rho_e).
\end{aligned}\tag{5.25}$$

Substituting (5.25) into (5.22) yields the following expression for the gradient:

$$\nabla J_e(\rho)=M_e(\rho)(\rho-\rho_e)\tag{5.26}$$

$$\begin{aligned}
M_e(\rho)&=\frac{\sigma_e^2}{\pi}\int_{-\pi}^{\pi}|H(e^{j\omega})|^2|G(e^{j\omega})|^2|S(e^{j\omega},\rho)|^2\\
&\quad\times\Re\{S_e^*(e^{j\omega})S(e^{j\omega},\rho)\bar{C}(e^{j\omega})\bar{C}^*(e^{j\omega})\}d\omega.
\end{aligned}$$

The similarity between (5.26) and (5.13) is obvious. The one relevant difference between these two expressions is the role of the reference spectrum $\Phi_r(e^{j\omega})$. In the matrix $M_e(\rho)$ appearing in (5.26), the noise spectrum appears *in lieu* of the reference spectrum present in matrix $M(\rho)$ of (5.13). Because the noise consists of filtered white noise, its spectrum is never zero (except perhaps at a finite number of frequencies represented by zeros of the transfer function $H(z)$ on the unit circle). So, there is no need to impose a minimum signal richness as an additional condition for the symmetric part of $M_e(\rho)$ to be positive definite—the noise is always sufficiently rich of infinite order. As a consequence, a similar result to Theorem 5.1 applies to the noise rejection objective function, but without the sufficiently rich condition in its statement.

Theorem 5.2 *Let Assumptions* B_y, LP *and* LI *be satisfied and let* $\Upsilon \subseteq \Gamma$ *be a connected set such that* $\rho_e \in \Upsilon$ *and that, for all* $\rho \in \Upsilon$,

$$\Re\left\{\frac{S_e(e^{j\omega})}{S(e^{j\omega},\rho)}\right\} > 0 \quad \forall \rho \in \Upsilon. \tag{5.27}$$

Then

$$(\rho - \rho_e)^T \nabla J_e(\rho) > 0 \quad \forall \rho \in \Upsilon, \rho \neq \rho_e.$$

All the remarks previously made about the reference tracking criterion $J_y(\rho)$ in Sect. 5.1 apply *ipsis literis* here, apart from those regarding the role of the reference spectrum.

Example 5.6 Let

$$G(z) = \frac{1}{z-0.5}, \qquad H(z) = \frac{z^2+z+0.45}{z^2-1.4z+0.45}, \qquad C(z,\rho) = \rho\frac{z}{z-0.9}. \tag{5.28}$$

It is easy to check that Assumption B_e is satisfied and that the ideal (minimum variance) controller, which minimizes $J_e(\rho)$, is achieved for $\rho_e = 2.4$, corresponding to the ideal closed-loop response given by

$$T_e(z) = \frac{2.4z}{z^2+z+0.45}.$$

The stability set is $\Gamma = (-0.05,\ 2.85)$. Because $T_e(z)$ in this example is the same as $T_d(z)$ in Example 5.4, and with the same process and controller structure, the analysis here is the same as in that example. It is straightforward to verify that $\frac{S_e(e^{j\omega})}{S(e^{j\omega},\rho)}$ is SPR for all $\rho \in \Upsilon = (1.35,\ 2.85)$. So, from Theorem 5.2, the cost $J_e(\rho)$ has no other extrema than ρ_e within the set Υ and this set is a candidate DOA for $J_e(\rho)$. For $\rho \notin \Upsilon$ the *SPR* condition is not satisfied.

5.3 The Mismatched Case

Let us analyze now the "mismatched control class" case, that is, the case where the ideal controller does not belong to the control class. An analysis for the reference tracking criterion $J_y(\rho)$ is presented, with similar results applying to the objective function $J_e(\rho)$. We will show that the convergence properties presented previously are robust with respect to the violation of Assumption B_y, that is, that these properties are maintained for moderate violations of this assumption.

One of the first consequences of Assumption B_y that we have seen in Sect. 5.1 and that was illustrated in Example 5.1 is that it implies that the optimization problem at our hands is well posed—that is, a global minimum actually exists. Now Assumption B_y is abandoned, but we still need a global minimum to exist in order for the design problem to make sense. So, we will assume explicitly that this is the case, and this assumption will replace Assumption B_y in our analysis.

Assumption C_y The objective function $J_y(\rho)$ presents a global minimum ρ_*, that is, $\exists \rho_* \in \Gamma: \ J_y(\rho) \geq J_y(\rho_*) \ \forall \rho \neq \rho_*$; moreover, the global minimum satisfies

$$\nabla J_y(\rho_*) = \mathbf{0}, \qquad \nabla^2 J_y(\rho_*) \geq 0.$$

Notice that Assumption C_y does not involve the existence of a *unique* global minimum; several global minima are admitted, maybe even an infinite number of them. Uniqueness of the global minimum requires signal richness as a necessary condition even in the mismatched case.

When Assumption B_y is satisfied, ρ_d is a global minimum ($\rho_* = \rho_d$) and $C(z, \rho_*) = C_d(z)$—the ideal controller can be achieved. When Assumption B_y is not satisfied, the best controller that can be obtained is $C(z, \rho_*) \neq C_d(z)$ and Assumption C_y replaces Assumption B_y in the analysis. The mismatch between the best controller allowed by the controller class under consideration and the ideal controller is defined as in (2.12) by the following transfer function

$$K(z) = C_d(z) - C(z, \rho_*). \tag{5.29}$$

Now substitute (5.29) into (5.9) to get

$$\begin{aligned}
\nabla J_y(\rho) = {} & \frac{1}{\pi} \int_{-\pi}^{\pi} \Phi_r(e^{\jmath\omega}) |G(e^{\jmath\omega}) S(e^{\jmath\omega}, \rho)|^2 \\
& \times \Re\{(C(e^{\jmath\omega}, \rho) - C(e^{\jmath\omega}, \rho_*) - K(e^{\jmath\omega}))^* S_d^*(e^{\jmath\omega}) S(e^{\jmath\omega}, \rho) \bar{C}(e^{\jmath\omega})\} d\omega \\
= {} & M(\rho)(\rho - \rho_*) - m(\rho)
\end{aligned} \tag{5.30}$$

where $M(\rho)$ is as defined previously in (5.15) and we have also defined a mismatch term $m(\rho)$:

$$m(\rho) = \frac{1}{\pi} \int_{-\pi}^{\pi} \Phi_r(e^{\jmath\omega}) |G(e^{\jmath\omega}) S(e^{\jmath\omega}, \rho)|^2 \Re\{S_d^*(e^{\jmath\omega}) S(e^{\jmath\omega}, \rho) \bar{C}(e^{\jmath\omega}) K^*(e^{\jmath\omega})\} d\omega. \tag{5.31}$$

Equation (5.30) is similar to (5.13), but perturbed by the vector function $m(\rho)$. This perturbation is unknown, continuous and satisfies $m(\rho_*) = \mathbf{0}$. In addition, $m(\rho)$ is bounded for all $\rho \in \Gamma$. Thus, $\|m(\rho)\|$ can be linearly bounded, that is, for any given set $\Upsilon \subseteq \Gamma$ containing the global optimum ρ_*:

$$\exists \alpha_\Upsilon \in \mathbb{R}^+: \quad \|m(\rho)\| < \alpha_\Upsilon \|\rho - \rho_*\| \quad \forall \rho \in \Upsilon \setminus \rho_*. \tag{5.32}$$

These properties of the mismatch term $m(\rho)$ allow the enunciation of the following theorem.

Theorem 5.3 *Let Assumptions* LP *and* LI *be satisfied,* $r(t)$ *be SRp and* α_Υ *be defined as in* (5.32). *Let* $\Upsilon \subseteq \Gamma$ *be a connected set such that* $\rho_* \in \Upsilon$ *and, for all* $\rho \in \Upsilon$:

$$\Re\{S_d^*(e^{\jmath\omega}) S(e^{\jmath\omega}, \rho)\} > 0 \quad \forall \omega.$$

If, in addition, the perturbation term $m(\rho)$ is such that its bound α_Υ in (5.32) satisfies $M_s(\rho) - \alpha_\Upsilon I_p > 0\ \forall \rho \in \Upsilon$, then

$$(\rho - \rho_*)^T \nabla J_y(\rho) > 0 \quad \forall \rho \in \Upsilon, \rho \neq \rho_*.$$

Proof Using (5.30) we have

$$(\rho - \rho_*)^T \nabla J_y(\rho) = (\rho - \rho_*)^T [M(\rho)(\rho - \rho_*) - m(\rho)]. \tag{5.33}$$

The theorem is proven by simple substitution of the assumptions in its statement into (5.33). □

Example 5.7 Let

$$G(z) = \frac{1}{z - 0.5}$$

and consider a reference tracking performance criterion with

$$T_d(z) = \frac{0.3}{z - 0.7};$$

then

$$C_d(z) = \frac{0.3z - 0.15}{z - 1}.$$

Consider the class of all delay-free integral controllers:

$$\mathcal{C} = \left\{ C(z) : C(z) = \rho \bar{C}(z),\ \rho \in \mathbb{R},\ \bar{C}(z) = \frac{z}{z - 1} \right\};$$

then Assumption B_y is not satisfied, that is, $C_d(z) \notin \mathcal{C}$.

The SPR condition (5.19) yields

$$\frac{S_d(z)}{S(z, \rho)} = \frac{z^2 + (\rho - 1.5)z + 0.5}{(z - 0.7)(z - 0.5)}$$

which is SPR for all $0 < \rho < 1.1$. This implies that this interval is a candidate DOA provided that $K(z)$—and hence $m(\rho)$—is small enough. Figure 5.8 shows a plot of the corresponding objective function for a reference consisting of a square wave such that the global minimum is at $\rho_* = 0.15$ (notice that because Assumption B_y is not satisfied, the global minimum depends on the spectrum of the reference). It is verified that the objective function has no other extrema in the predicted interval.

Consider now the same process and reference model but a different class of controllers, described by

$$\mathcal{C} = \left\{ C(z) = \frac{\rho z + 0.3}{z - 1}; \rho \in \mathbb{R} \right\};$$

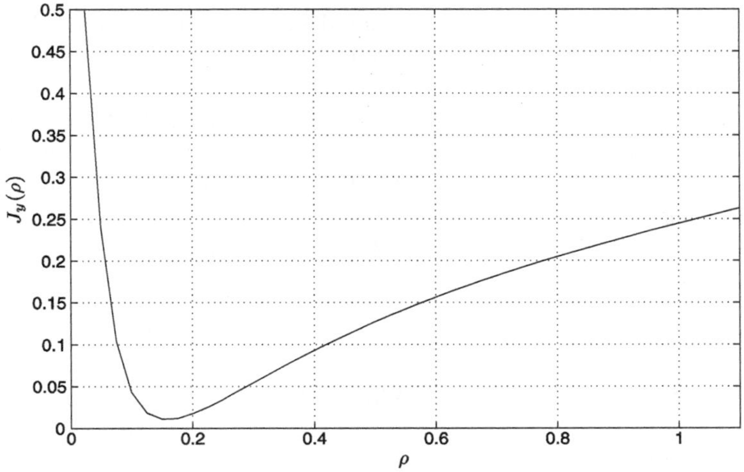

Fig. 5.8 Example 5.7—reference tracking criterion

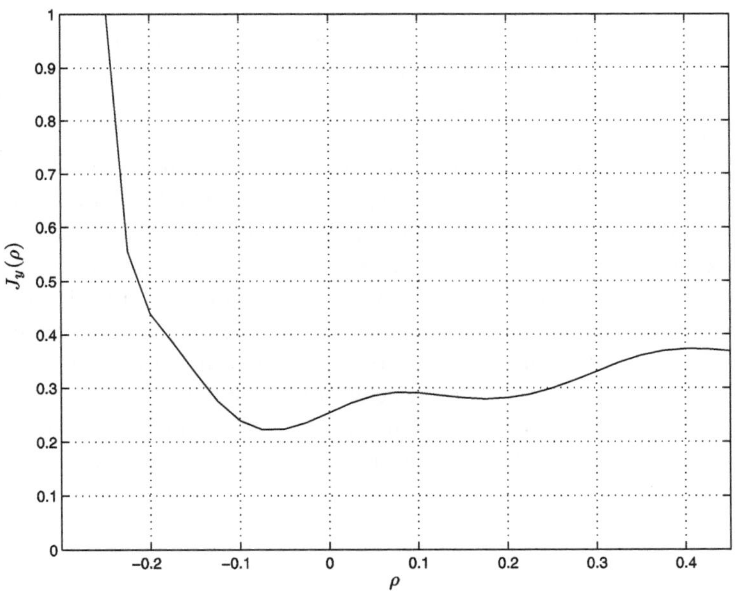

Fig. 5.9 Example 5.7—reference tracking criterion as a function of ρ

then

$$\frac{S_d(z)}{S(z,\rho)} = \frac{z^2 + (\rho - 1.5)z - 0.2}{(z - 0.7)(z - 0.5)},$$

which is SPR in the interval $-0.3 < \rho < 0.44$. However, the cost function is as presented in Fig. 5.9, where it can be seen that this interval is not a candidate DOA.

The nice properties of the cost function are destroyed by this additional distance between the ideal controller and the controller class.

The "nice" properties of the objective function, obtained for the case where Assumption B_y is satisfied, are nevertheless robust to the violation of this hypothesis. Even in the mismatched control case the SPR condition still serves as a guide to determine the appropriate initializations, provided that the mismatch between the ideal controller and the best controller that can be obtained within the class of controllers available is moderate.

5.4 Choosing the Algorithm Parameters

The optimization of the objective function will be performed, as discussed previously, by an algorithm of the form

$$\rho_{i+1} = \rho_i - \gamma_i R_i \nabla J(\rho_i) \tag{5.34}$$

with $R_i \in \mathbb{R}^{p\times p}$ and $\gamma_i \in \mathbb{R}^+$. So, there are two parameters to be specified by the user and that will determine whether or not the algorithm will converge and, if it does, at what rate: the search directions at each step, defined by the matrices R_i, and the sequence of step sizes γ_i.

Among the myriad of choices existent for the search direction, we have chosen in the previous chapter to analyze the two most widely known in general and most commonly applied in data-driven control design: the steepest descent and the Newton-Raphson algorithms. These two were presented in some detail and their convergence features were analyzed, an analysis that will guide the brief discussion of the choice of the search directions that will be presented next. Once the search direction policy is fixed, the relevant properties of the algorithm (5.34) are explored to design a step size sequence that ensures convergence to the global minimum (when the given initial condition is such that this is possible) as quickly as possible. The analysis and the corresponding design of the step size sequence are valid for any search direction.

5.4.1 The Search Direction

The steepest descent algorithm corresponds to the algorithm (5.34) with $R_i = I_p$, whereas the Newton-Raphson's algorithm corresponds to $R_i = (\nabla^2 J(\rho_i))^{-1}$ and $\gamma_i = 1\ \forall i$. A slight adaptation of the Newton-Raphson algorithm is to keep its direction but allow different step sizes, which keeps the fundamental properties of the Newton-Raphson search direction allowing additional flexibility that can be used to further improve the method's convergence. This adaptation will be accordingly named the variable step Newton-Raphson algorithm.

It was shown in the previous chapter that the Newton-Raphson algorithm typically results in higher convergence rates and the steepest descent algorithm presents larger domains of attraction. Then from a given initial condition that is potentially far from the optimum, it is safer to use the steepest descent algorithm. But from a better initial condition, it is smarter to use the Newton-Raphson algorithm, as it will converge faster. It is then only logical to use the steepest descent algorithm in the initial steps of the optimization, when the parameter approximation is farther away from the optimum, switching to the Newton-Raphson algorithm once the parameter approximation is sufficiently close to it. This is actually the search direction policy that we prefer to apply and recommend.

5.4.2 The Step Sizes—First Solution

The choice of γ_i is fundamental to the performance of the algorithms. If too small steps are taken at each iteration, then convergence will be very slow. Large values are dangerous because they can cause the algorithm to leave the candidate DOA and even result in an unstable closed-loop. Among the many possible solutions for the choice, let us present two that have proven quite effective in data-driven control design.

The first one is a recursive step size policy, in which the steps are maintained fixed as long as the cost function is reduced, then reduced by half when the cost function is increased at a given iteration. This step size sequence results in convergence to the global minimum under quite general conditions, as shown below.

Theorem 5.4 *Let ρ_* be the global minimum of the cost function $J(\rho):\mathbb{R}^p \to \mathbb{R}^+$. Assume that there exists $\epsilon > 0$ and $a > 0$ such that $\rho_0 \in \mathcal{B}_\epsilon(\rho_*)$ and*

$$\frac{(\rho-\rho_*)^T R_i \nabla J(\rho)}{\|\rho-\rho_*\|\|R_i \nabla J(\rho)\|} > a > 0 \quad \forall \rho \in \mathcal{B}_\epsilon(\rho_*) \setminus \rho_*.$$

The algorithm (5.34) *converges to ρ_* if the step size sequence is given by*:

$$\gamma_i = \frac{\mu_i}{\|R_i \nabla J(\rho_i)\|}$$

where μ_i is chosen as:

- $\mu_i = \mu_{i-1}$ *if* $J_i(\rho) \le J_{i-1}(\rho)$;
- $\mu_i = \frac{\mu_{i-1}}{2}$ *if* $J_i(\rho) > J_{i-1}(\rho)$.

Proof Consider the quadratic candidate Lyapunov function

$$V(\rho) = (\rho-\rho_*)^T(\rho-\rho_*) = \|\rho-\rho_*\|^2.$$

Let $\mu_0 = \mu$ be such that $V(\rho_1) < V(\rho_0)$. This is no loss of generality, because if this were not the case then the step size policy would require to reduce the step until such a μ is found. For such a μ:

$$
\begin{aligned}
V(\rho_{i+1}) - V(\rho_i) &= \|\rho_{i+1} - \rho_*\|^2 - \|\rho_i - \rho_*\|^2 \\
&= \left\| \rho_i - \mu \frac{R_i \nabla J(\rho)}{\|R_i \nabla J(\rho)\|} - \rho_* \right\|^2 - \|\rho_i - \rho_*\|^2 \\
&= \mu^2 - 2\mu \frac{(\rho_i - \rho_*)^T R_i \nabla J(\rho)}{\|R_i \nabla J(\rho)\|} \\
&< \mu^2 - 2\mu a \|\rho_i - \rho_*\|
\end{aligned}
$$

as long as the Lyapunov function keeps being reduced at each iteration, because in this case μ_i is not altered. But this implies that for any ρ_i satisfying

$$
\|\rho_i - \rho_*\| > \frac{\mu}{a}
$$

the Lyapunov function is reduced at least by a certain fixed amount:

$$
V(\rho_{i+1}) - V(\rho_i) < -\mu^2.
$$

That is, as long as ρ_i is outside a ball $\mathcal{B}_\epsilon(\rho_*)$ with $\epsilon = (\frac{\mu}{a})^2$, the Lyapunov function is reduced of at least $-\mu^2$ at each iteration. This implies that the parameter ρ will enter the ball $\mathcal{B}_\epsilon(\rho_*)$ in finite time. More specifically, let k be the first integer larger than $\frac{V(\rho_0)}{\mu^2} - \frac{1}{a^2}$; then the parameter ρ enters $\mathcal{B}_\epsilon(\rho_*)$ in at most k iterations.

Once the algorithm has entered this ball, the Lyapunov function will be increased. When this happens, the parameter μ_i will be reduced by half repeatedly until reaching a value—say μ'—such that the Lyapunov function is reduced. Then the same argument above shows that the parameter enters in finite time a ball $\mathcal{B}_\epsilon(\rho_*)$ with $\epsilon = (\frac{\mu'}{a})^2$, whose radius is at most half the one of the previous ball. As the iterations pass, the radii of the balls entered by the algorithm in finite time keep decreasing, causing the parameter to converge to the global minimum. □

This is a recursive choice for the step sizes, where each γ_i is chosen as a function of the previous one. So, it is necessary to initialize the step size, which can be done by establishing the size of the first parameter change as a fraction of the initial parameter value:

$$
\gamma_0 = \alpha \|\rho_0\| \frac{1}{\|R_0 \nabla J(\rho_0)\|},
$$

and the fraction α is chosen to make a first step which the designer deems as a safe change to be made.

5.4.3 The Step Sizes—Second Solution

It has been shown in Chap. 4 that the algorithm (5.34) converges to the global optimum if

$$
-2\gamma_i(\rho_i - \rho_*)^T R_i \nabla J(\rho_i) + \gamma_i^2 \nabla J(\rho_i)^T R_i^T R_i \nabla J(\rho_i) + \omega(\|\rho_i - \rho_*\|) < 0
$$

at all iterations. Using the expression $\nabla J(\rho_i) = M(\rho_i)(\rho_i - \rho_*)$ for the gradient of H_2 cost functions and the class K function $\omega(\|\rho_i - \rho_*\|) = a^2\|\rho_i - \rho_*\|^2$, the above condition can be expressed as

$$-2\gamma_i(\rho_i - \rho_*)^T R_i M(\rho)(\rho_i - \rho_*) + \gamma_i^2(M(\rho)(\rho_i - \rho_*))^T R_i^T R_i M(\rho)(\rho_i - \rho_*) + a^2\|\rho_i - \rho_*\|^2 < 0$$

which can be simplified to

$$(\rho_i - \rho_*)^T(-2\gamma_i R_i M(\rho) + \gamma_i^2 M^T(\rho) R_i^T R_i M(\rho) + a^2)(\rho_i - \rho_*) < 0.$$

This is a quadratic form which is satisfied if

$$-2\gamma_i R_i M(\rho_i) + \gamma_i^2 M^T(\rho_i) R_i^T R_i M(\rho_i) + a^2 < 0. \tag{5.35}$$

Using the Schur's Lemma we obtain the equivalent condition

$$\begin{bmatrix} 2\gamma_i R_i M(\rho_i) - a^2 & \gamma_i R_i M(\rho_i) \\ \gamma_i M^T(\rho_i) R_i^T & I \end{bmatrix} > 0. \tag{5.36}$$

Then we can enunciate the following theorem, which is valid for arbitrary search directions.

Theorem 5.5 *Suppose that Assumption* B_y *is satisfied and let* $\mathcal{B}_\alpha(\rho_d) \subseteq \Gamma$ *be a candidate DOA. If* $\rho_0 \in \mathcal{B}_\alpha(\rho_d)$ *and* γ_i *respects* (5.36) *for some* $a > 0$ *then* $\lim_{i\to\infty} \rho_i = \rho_d$.

It is important to realize that $\mathcal{B}_\alpha(\rho_d)$ being a candidate DOA means that there is a sequence of γ_i that satisfies the conditions of the theorem (see Definition 4.8). But besides finding some sequence that results in convergence, we would like to find the step size sequence that also results in the fastest convergence. The convergence rate of the algorithm is directly related to the parameter a; for larger values of a the algorithm converges faster. So, to guarantee convergence to the global minimum ρ_d and to make this convergence as fast as possible, the step size γ_i must be chosen as the solution of the following problem

$$\max_{a,\gamma_i} a$$

$$\text{subject to (5.36)}$$

An important feature is that such a selection of the step size can be made automatically, and is based solely on the basis of R_i and an estimate of $M(\rho)$. This second method uses more information about the process (an estimate of $M(\rho)$) than the first method to compute the step size sequence.

5.4.3.1 Estimate of $M(\rho)$

In this section we discuss how to obtain from the data the information required for computation of the step sizes γ_i. This computation requires an estimate of the matrix $M(\rho)$, which can be expressed as

$$M(\rho) = \sum_{i=1}^{n_r} \Phi_r |G|^2 |S(\rho)|^2 \Re\{S_d^* S(\rho) \bar{C} \bar{C}^*\} \tag{5.37}$$

when the reference has a finite support with n_r components. The matrix $M(\rho)$ depends on the reference signal, on the controller, on the reference model and on the frequency response of the process. The reference signal, the controller and the reference model are known by the designer and the only information unknown to compute $M(\rho)$ concerns the frequency response of the process. A rough estimate of the process response can be made using the same input-output data collected to estimate the gradient of the cost function. Since the quality of this estimate is not critical for the convergence of the algorithm, using an estimate of the steady-state gain and of the pass band as representative of this frequency response is appropriate to approximate the sum (5.37). This procedure is successfully applied in some examples along this book, including the experimental results presented in Chap. 8.

5.5 A Case Study

Consider the control of a process described by (there is no noise):

$$y(t) = \frac{0.05}{z - 0.95} u(t). \tag{5.38}$$

This process is controlled by a PI controller

$$C(z, \rho) = \begin{bmatrix} \varrho_1 & \varrho_2 \end{bmatrix} \begin{bmatrix} \frac{z}{z-1} \\ \frac{1}{z-1} \end{bmatrix}. \tag{5.39}$$

The process' transfer function is unknown to the designer, whom will be applying a data-driven design to the PI controller. It is only assumed that the designer knows that the process' transfer function is of first order. With this information, the designer can choose his/her reference model so that Assumption B_y can be verified. The following reference model is specified:

$$T_d(z) = \frac{0.1}{z - 0.9}.$$

The above reference model can be achieved by the ideal controller

$$C_d(z) = \begin{bmatrix} 2 & -1.9 \end{bmatrix} \begin{bmatrix} \frac{z}{z-1} \\ \frac{1}{z-1} \end{bmatrix}. \tag{5.40}$$

This controller lies in the class of PI controllers in the form (5.39), so Assumption B_y is satisfied.

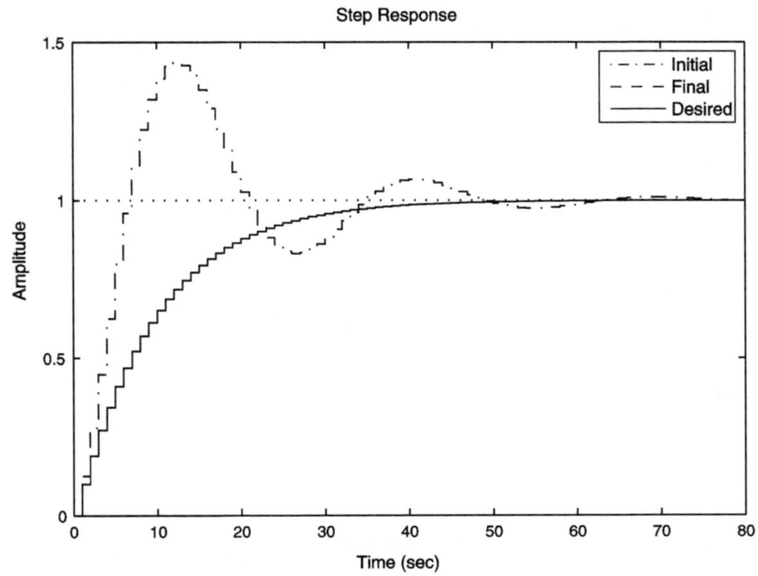

Fig. 5.10 Closed-loop step response: desired response, with the initial controller, and with the final controller (hidden by the desired response)

Consider that the system is initially in closed-loop with the controller

$$C(z, \rho_0) = \begin{bmatrix} 2.5 & -1.5 \end{bmatrix} \begin{bmatrix} \frac{z}{z-1} \\ \frac{1}{z-1} \end{bmatrix}. \tag{5.41}$$

Figure 5.10 shows the step response of the closed-loop system. This initial response is very different from the desired one, as can be seen visually and assessed by the significant value of the objective function $J(\rho_0) = 0.15048$: given that the input is a unit step, this is roughly a 15% error in the desired value for the output.

Let us use the algorithm (5.34)[2] to improve the closed-loop performance. The following reference signal, which guarantees persistence of excitation, is used to obtain the data:

$$r(t) = sq\left(\frac{2\pi t}{200}\right)$$

where $sq(\frac{2\pi t}{T})$ stands for a square wave with period T.

The search direction is initially the gradient direction—the steepest descent method $R_i = I_p$—switching to the variable step Newton-Raphson after the parameter value is sufficiently close to the minimum. The step sizes are determined by the second solution presented earlier in Sect. 5.4.3. The result of this iterative design is presented in Fig. 5.11 and in Table 5.1, where the evolution of the step sizes,

---

[2]The gradient $\nabla J(\rho)$ was estimated based on data using the Iterative Feedback Tuning (IFT) method described in Sect. 7.1.

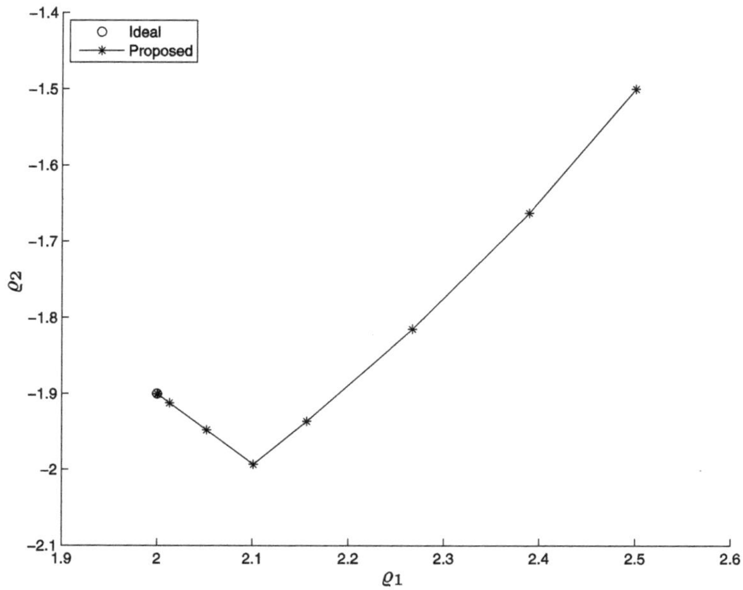

Fig. 5.11 Evolution of the controller parameters using different search directions and the step size sequence of Sect. 5.4.3

Table 5.1 Controller parameters evolution

i	J_i	ρ_i^T	γ_i	R_i
0	0.15048	[2.5000 −1.5000]	1.3537	Identity
1	0.11803	[2.3887 −1.6630]	1.0351	Identity
2	0.07437	[2.2670 −1.8149]	0.6123	Identity
3	0.02267	[2.1565 −1.9366]	0.2267	Identity
4	0.00031	[2.1012 −1.9931]	0.4916	Hessian
5	$7.8514e-005$	[2.0518 −1.9481]	0.7472	Hessian
6	$4.4729e-006$	[2.0132 −1.9123]	0.9421	Hessian
7	$1.2461e-008$	[2.0008 −1.9007]	0.9955	Hessian
8	$2.1745e-013$	[2.0000 −1.9000]	0.9958	Hessian

the search directions, the controller parameters and the resulting values of the cost function are all presented. It is seen that the minimization is successful, obtaining five correct significant digits of ρ_d in few iterations. The step size has converged to a value very close to one, which amounts to the original Newton-Raphson method; when very close to the minimum the Newton-Raphson is indeed the optimal algorithm. Figure 5.10 also shows the step response with the final controller (iteration 8), which is visually indistinguishable from the reference model response.

To further stress the effectiveness of this particular design, let us present another, more classical alternative, in which the steepest descent algorithm is applied with step sizes originated from the classical stochastic approximation theorem, as given in (4.22). It is worth of notice that this "classical" step size policy does not provide

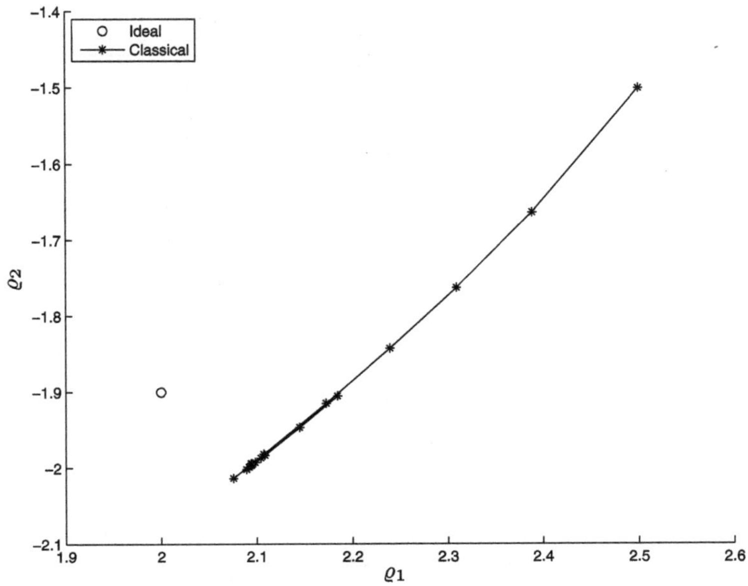

Fig. 5.12 Evolution of the controller parameters with the steepest descent algorithm and the classical step size sequence given by the harmonic sequence

an initial step size, which is then an additional parameter for the designer to worry about. To make an easier comparison, we chose the step size of the first iteration of the classical choice the same as calculated before. The result of this alternative optimization is presented in Fig. 5.12 and in Table 5.2. It is clear from the table that convergence is slower and is not monotonic, and it is apparent in the figure that the steepest descent gets trapped after some iterations. This is not because of a local minimum of the cost function, but rather because the component of the gradient of the cost function along the line connecting this point to the global minimum is very small. It was the switching to the Newton-Raphson's search direction that allowed the minimization to evolve along this line and converge to the global minimum in the previous case.

From these results, one may be tempted to apply the Newton-Raphson algorithm from the first step; it turns out that this is not a good idea because of the Newton-Raphson's small DOA. Indeed, Table 5.3 and Fig. 5.13 show the result of applying the Newton-Raphson method with the initial condition (5.41). In this case, because the initial condition is far from the global minimum, the Newton-Raphson method diverges, arriving at an unstable closed-loop at iteration 2.

5.6 Chapter Conclusions

Most developments, results and conclusions in this chapter were given for the reference tracking performance criterion $J_y(\rho)$. These properties are a consequence of

Table 5.2 Evolution of the controller parameters with the steepest descent and the classical step sizes

i	J_i	ρ_i^T	γ_i	R_i
0	0.15048	[2.5000 −1.5000]	1.35370	Identity
1	0.11803	[2.3887 −1.6630]	0.67685	Identity
2	0.09127	[2.3091 −1.7623]	0.45123	Identity
3	0.06350	[2.2393 −1.8423]	0.33842	Identity
4	0.03202	[2.1724 −1.9145]	0.27074	Identity
5	0.00214	[2.1074 −1.9816]	0.22562	Identity
6	0.00886	[2.0754 −2.0133]	0.19339	Identity
7	0.03721	[2.1843 −1.9049]	0.16921	Identity
8	0.01740	[2.1446 −1.9462]	0.15041	Identity
9	0.00207	[2.1083 −1.9830]	0.13537	Identity
10	0.00078	[2.0894 −2.0018]	0.12306	Identity
11	0.00111	[2.1043 −1.9867]	0.11281	Identity
12	0.00027	[2.0920 −1.9988]	0.10413	Identity
13	0.00020	[2.0982 −1.9924]	0.09663	Identity
14	0.00010	[2.0940 −1.9964]	0.09027	Identity
19	$7.9126e-005$	[2.0949 −1.9949]	0.06765	Identity
29	$7.8385e-005$	[2.0945 −1.9944]	0.04513	Identity
39	$7.7867e-005$	[2.0941 −1.9941]	0.03382	Identity
49	$7.7468e-005$	[2.0939 −1.9939]	0.02704	Identity

Table 5.3 Evolution of the controller parameters with the Newton-Raphson method

i	$J(\rho_i)$	ρ_i^T	γ_i	R_i
0	0.15048	[2.5000 −1.5000]	1	Hessian
1	0.09372	[4.2419 −3.5074]	1	Hessian
2	∞	[5.1033 −5.3243]	1	Hessian

the particular structure of the gradient in the form (5.13). Since the noise rejection performance criterion $J_e(\rho)$ presents this same structure, as seen in (5.26), the same developments, results and conclusions apply to this performance criterion as well, apart from the ones related to signal richness.

The convergence analysis is based on the positivity of the matrix $M_s(\rho)$ in (5.15) (or the symmetric part of $M_e(\rho)$ in (5.26) for the noise rejection performance criterion). This positivity condition can be used in many different ways in a data-driven design. For example, whenever the symmetric matrix $M_s(\rho_i)$ is positive definite at some ρ_i, the next iteration will drive the parameter closer to the global optimum. So this condition can be used as a test to be carried out at each iteration to verify whether or not the algorithm is converging to the global minimum. It is also possible to verify *a priori* the positivity of this matrix for all parameter values in a given ball, thus establishing that this ball is a candidate DOA (or not). Moreover, when the initial controller is within a candidate DOA, then this same condition provides an appropriate automatic procedure to calculate the step sizes. We have presented

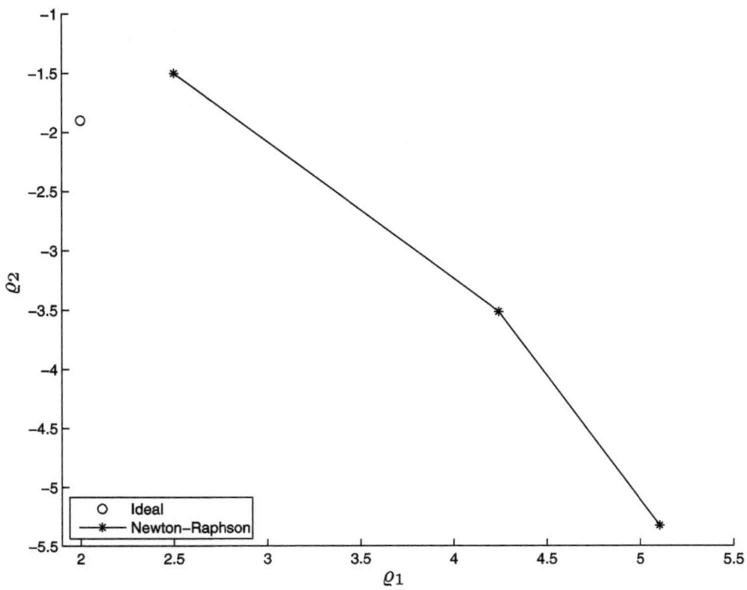

Fig. 5.13 Evolution of the controller parameters with the Newton-Raphson algorithm

an example suggesting that much faster convergence can be expected with the step sizes calculated in this way when compared to other, more classical alternatives. This expectation is also supported by other examples given elsewhere (see, for instance, [1]).

A conceptually and practically relevant alternative convergence condition is that a particular transfer function, which is the ratio between two sensitivities, is SPR. This is a more conservative condition than the positivity of $M_s(\rho)$, but it does not depend on the reference spectrum and provides important insight to the problem. Similar SPR conditions abound in the literature of system identification, adaptive control and data-driven control. Here we have obtained this condition as a general property of data-driven and adaptive control design, analyzing only the particularities of the H_2 cost functions. Even though our analysis has been presented for the steepest descent algorithm, for other search directions the same arguments will apply with $M_s(\rho_i)$ replaced by $R_i M_s(\rho_i)$.

On the other hand, instead of verifying candidate DOAs for an optimization problem fully defined *a priori*, one can shape the problem such that given sets are candidate DOAs, or that given parameter values are within candidate DOAs. How to do this is the subject of the next chapter.

References

1. D. Eckhard, A.S. Bazanella, Data-based controller tuning: Improving the convergence rate, in *Decision and Control (CDC), 2010 49th IEEE Conference on*, (2010), pp. 4801–4806

Chapter 6
Cost Function Shaping

6.1 Introduction

The previous chapter presented an analysis of the properties of the cost function around the global minimum and we have seen conditions under which the cost function is "well behaved". Having a "good behavior" means that large candidate domains of attraction exist, without extrema close to the global minimum that would complicate or even preclude the optimization. In this chapter, the same problem is studied from a synthetic perspective, that is, instead of analyzing the properties of a given problem, we will see how the designer can choose some parameters of the problem to make the cost function "well behaved". A set of principles and procedures to achieve this goal will be derived, always based on the analytical results in the previous chapter, and the ensemble of these tools has been baptized *cost function shaping*.

6.2 The Problem Data

For each performance criterion, satisfaction of the basic convergence condition (4.12) is determined through the positivity of a particular symmetric real matrix. For the reference tracking criterion $J_y(\rho)$ the relevant matrix is $M_s(\rho)$ given in (5.15), whereas for noise rejection $J_e(\rho)$ the relevant matrix is the symmetric part of $M_e(\rho)$ given in (5.26). Among the variables present in (5.15) and (5.26), which define the corresponding matrix and therefore determine the convergence properties, some can be manipulated by the designer and others cannot. These variables are the following, divided into three groups:

- the process characteristics—$G(z)$, $H(z)$ and σ_e^2;
- the controller features—the controller class $\mathcal{C}$, its parameterization $\bar{C}(z)$ and the initial controller parameter ρ_0;
- the performance criterion—$T_d(z)$ and Φ_r.

A. Sanfelice Bazanella et al., *Data-Driven Controller Design*,
Communications and Control Engineering,
DOI 10.1007/978-94-007-2300-9_6,

The first group of variables are the process characteristics, which obviously are given and certainly cannot be changed. Moreover, they are unknown but at least some partial information about them is provided—such as the relative degree of $G(z)$, or an estimate of σ_e^2, for example.

The second group of variables concerns the controller. The controller class $\mathcal{C}$ is given and known; it is a designer's choice that is usually made in a previous stage of system conception, so in most cases we cannot change it at this stage. The hardware and software available often impose the controller class, linearly parameterized PID controllers being the most commonly used. On the other hand, the particular parametrization $\bar{C}(z)$ used to represent this class can often be manipulated. It has been proven earlier in this book that this parametrization must represent the class $\mathcal{C}$ with the minimal number of parameters but it can otherwise be freely chosen. This choice does play a role in the properties of the cost function, and thus might be useful for cost function shaping.

The initial controller parameter ρ_0 could, in theory, be selected at will. However, finding an initial controller that is guaranteed to provide a stable closed loop is not without danger when the process is unknown. Also, a major application field for model-free tuning methods is the performance improvement of controllers that have been operating in a stable, but not optimally performing way. In this situation, which is probably the most common in practice, the initial controller is the one already operating in the control loop and is thus imposed upon us. So, whether it is possible to choose ρ_0 in a given practical situation is determined by case-specific and rather subjective considerations.

The remaining group of variables influencing the objective function are related to the performance criterion: $T_d(z)$ and Φ_r. Once it has been established what is the desired performance criterion, these variables are fixed; changing them would imply minimizing another criterion, achieving a performance that is not the one desired. Nevertheless, if the criterion of choice is too hard to optimize starting from the starting controller ρ_0, we may consider minimizing an easier criterion as an intermediate task. Then, taking the new controller resulting from this optimization as the initial controller, it might be easier to optimize the desired criterion. Actually, one can think of inserting more than one intermediate task, optimizing at each time a criterion that is closer to the desired one, and guaranteeing that each one of these intermediate optimization tasks will converge to its global minimum. This cautious approach to performance improvement is a familiar idea that has been called "the windsurfer approach" to control design by some, or simply *cautious control* by many. A similar reasoning applies to the reference spectrum: if it is too hard to minimize the performance criterion for the desired reference, then minimize it for other easier to track references as intermediate steps, until the desired reference is tracked.[1]

This is the central idea presented in this chapter: one can manipulate one or both of the variables $T_d(z)$ and Φ_r stepwise so that the resulting intermediate cost functions have a larger domain of attraction to their global optimum, in such a way as

[1] We will see that under Assumption B_y this is not even necessary.

to eventually minimize the desired cost function. The effects and possibilities of manipulating each one of these variables are now studied to derive a number of different numerical maneuvers and experimental procedures that can be used to make the cost function to be optimized "well behaved". The ensemble of these maneuvers and procedures is what we have baptized "cost function shaping".

6.3 Cautious Control

Convergence to the global optimum is obtained if the optimization is initialized sufficiently close to it, where closeness is measured by a metric defined as the phase difference between the desired and achieved sensitivities, stemming from (5.17). So, choosing an initial controller ρ_0 such that this difference is sufficiently small should be appropriate. However, it is often the case that the controller to be tuned is in operation, that it is stabilizing the plant, but providing rather poor performance. The engineer may be very reluctant to change abruptly the controller parameters to another value at which not even stability is 100% assured—after all, one bird in the hand is worth two in the bush. So, instead of choosing an initial controller whose parameters are close to the optimal ones, let us consider an alternative: keep ρ_0 unchanged and temporarily change $T_d(z)$ so that the optimal parameter ρ_d is brought closer to ρ_0.

Starting from an initial controller which delivers a given performance—say $T_0(z)$—which is considered to be poor, let us choose a first intermediate reference model $T_d^1(z)$. This reference model should not require at once the achievement of a performance that is much better than the one we already have with $T_0(z)$, because this would likely require the optimal controller parameters to be too far from the current controller parameters. Instead, $T_d^1(z)$ should be *cautious*, aiming at a modest performance improvement, one which is closer to the (poor) performance $T_0(z)$ than the real reference model of interest—namely $T_d(z)$. Of course, closeness here should be understood as measured by the appropriate metric defined previously, that is, the distance

$$\text{dist}(\rho_0, \rho_d) = \max_{\omega} \angle \frac{S_d^1(e^{j\omega})}{S(e^{j\omega}, \rho_0)}$$

should be small, where $S_d^1(z) = 1 - T_d^1(z)$. Specifically, one should try to ascertain that $\frac{S_d^1(z)}{S(z,\rho_0)}$ is SPR.

Then an iterative procedure is applied to find the global minimum of this cautious intermediate criterion—say ρ_*^1. Once the global optimum of this intermediate criterion has been reached, we can pick a second, more ambitious, reference model $T_d^2(z)$ (i.e. one closer to $T_d(z)$), and optimize it starting from ρ_*^1 as the initial controller. This argument can be used successively, with several intermediate reference models, until the desired reference model $T_d(z)$ is achieved. We illustrate the procedure by means of an example.

Example 6.1 Consider again the following system and reference model:

$$G(z) = \frac{1}{z - 0.5} \qquad C(z, \rho) = \rho \frac{z}{z - 0.9} \qquad T_d(z) = \frac{2.4z}{z^2 + z + 0.45}$$

and let us say that the initial controller parameter is $\rho_0 = 0.4$. Then the closed-loop transfer function is given by

$$T(z, \rho_0) = \frac{0.4z}{z^2 - z + 0.45}.$$

Consider initially a reference tracking objective, where the control system must track a square wave reference with period $T = 6$ samples and unitary amplitude with the performance specified by $T_d(z)$.[2]

In order to verify what would happen if an iterative optimization of the cost function $J_y(\rho)$ were applied to the design of the controller, the cost function is calculated for all values of $\rho \in \Gamma$, resulting in the graphic shown in Fig. 6.1d. It is seen that this cost function presents a local maximum at $\rho \approx 0.6$. Under these circumstances, convergence to the global optimum $\rho_d = 2.4$ is not possible from the initial condition $\rho_0 = 0.4$. In fact, an attempt to apply a data-driven design in this case would likely be disastrous. The algorithm would reduce the controller parameter ρ indefinitely, which depending on the step sizes would either make it to assume values very close to zero (which would amount to no control at all), or cross the boundary of the stability set into negative values that would result in instability.

Let us see how cautious control would work here. Consider the use of a cautious reference model given by

$$T_d^1(z) = \frac{0.6z}{z^2 - 0.8z + 0.45},$$

for which $\rho_d^1 = 0.6$, and verify the behavior of the corresponding cost function

$$J_y^1(\rho) = \bar{E}[(T_d^1(z) - T(z, \rho))r(t)]^2.$$

This function is plotted in Fig. 6.1a, where it is seen that the initial controller $\rho_0 = 0.4$ belongs to a candidate DOA of the global minimum $\rho_d^1 = 0.6$. Optimization of the cost function $J_y^1(\rho)$ starting from ρ_0 is then "easy" and will result in reaching the global minimum $\rho_d^1 = 0.6$.

Once this optimization has been performed, we can try a second step, with the new intermediate reference model

$$T_d^2(z) = \frac{1.2z}{z^2 - 0.2z + 0.45},$$

[2]Granted, this is a rather artificial control objective, but the purpose of this example is not to illustrate practical applications but rather to explain the ideas. Accordingly, the choice of this particular example was based on obtaining easily understandable equations and pictures rather than on its practical meaning. Real life examples are given elsewhere, particularly in Chap. 8.

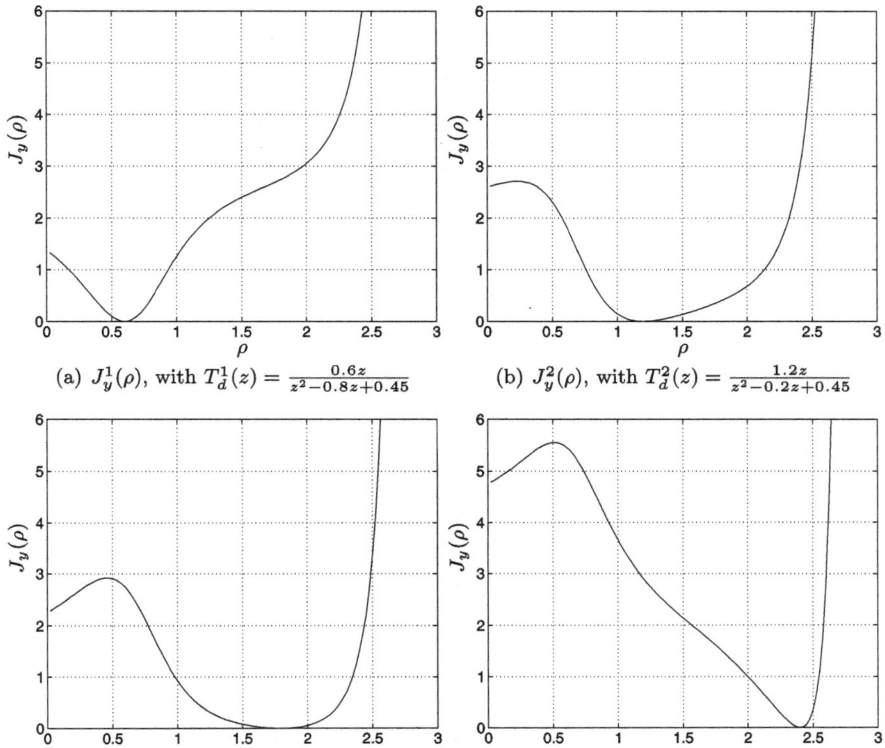

(a) $J_y^1(\rho)$, with $T_d^1(z) = \frac{0.6z}{z^2-0.8z+0.45}$ (b) $J_y^2(\rho)$, with $T_d^2(z) = \frac{1.2z}{z^2-0.2z+0.45}$

(c) $J_y^3(\rho)$, with $T_d^3(z) = \frac{1.8z}{z^2+0.4z+0.45}$ (d) $J_y(\rho)$, with $T_d(z) = \frac{2.4z}{z^2+z+0.45}$

Fig. 6.1 Example 6.1—cautious control: performance criterion $J_y(\rho)$ for square-wave reference, $T = 6$ s, with different reference models

for which $\rho_d^2 = 1.2$. The corresponding cost function $J_y^2(\rho)$ is plotted in Fig. 6.1b, and it is clear that ρ_d^1, which represents the initial condition for the optimization of $J_y^2(\rho)$, is within a candidate DOA of the global minimum $\rho_d^2 = 1.2$. Having done this second optimization, we take ρ_d^2 as the initial condition for the optimization of a third intermediate reference model

$$T_d^3(z) = \frac{1.8z}{z^2 + 0.4z + 0.45},$$

for which $\rho_d^3 = 1.8$; the corresponding cost function $J_y^3(\rho)$ is plotted in Fig. 6.1c. Finally, from ρ_d^3 as the initial condition it is safe to optimize the desired criterion $J_y(\rho)$ with the reference model

$$T_d(z) = \frac{2.4z}{z^2 + z + 0.45}.$$

As a matter of fact, the above sequence of reference models is more cautious than necessary, since the first optimum ρ_d^1 is already within a candidate DOA of ρ_d. But

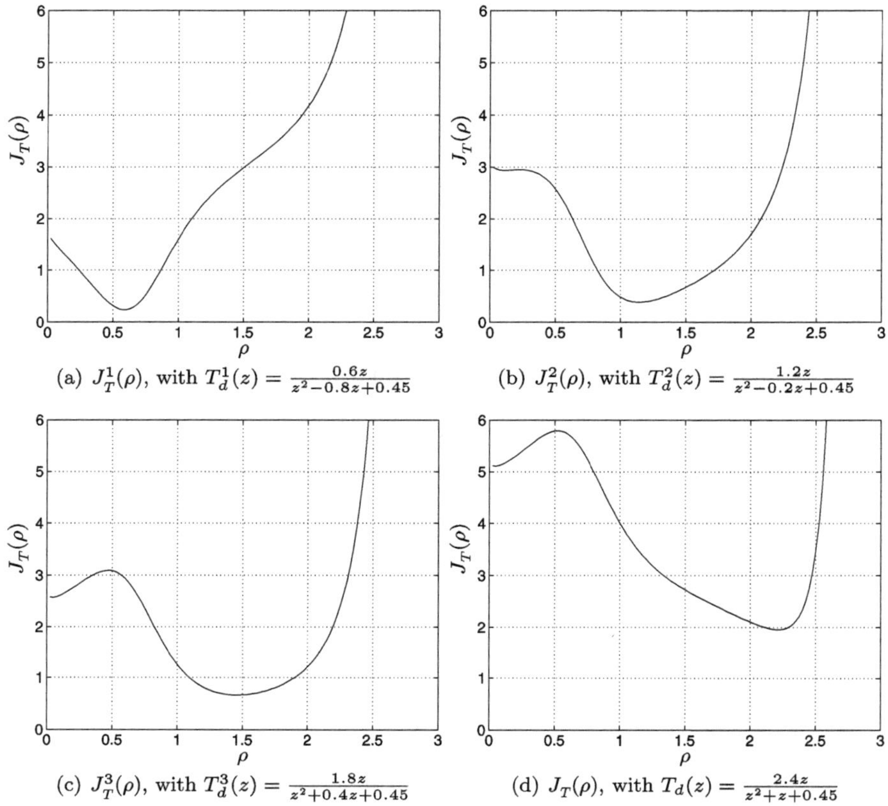

(a) $J_T^1(\rho)$, with $T_d^1(z) = \frac{0.6z}{z^2-0.8z+0.45}$

(b) $J_T^2(\rho)$, with $T_d^2(z) = \frac{1.2z}{z^2-0.2z+0.45}$

(c) $J_T^3(\rho)$, with $T_d^3(z) = \frac{1.8z}{z^2+0.4z+0.45}$

(d) $J_T(\rho)$, with $T_d(z) = \frac{2.4z}{z^2+z+0.45}$

Fig. 6.2 Example 6.1—cautious control: performance criterion $J_T(\rho)$ for square-wave reference, $T = 6$ s, $\sigma_e = 0.2$ and $H(z) = \frac{z^2-1.15z+0.45}{z^2-1.4z+0.45}$, with different reference models

in an actual application we would not know that, so going directly from ρ_d^1 could represent a relevant risk.

Consider now that the performance criterion is no longer reference tracking $J_y(\rho)$, but the composite criterion $J_T(\rho) = J_y(\rho) + J_e(\rho)$, with the same reference signal, $\sigma_e = 0.2$ and $H(z) = \frac{z^2-1.15z+0.45}{z^2-1.4z+0.45}$. The concept of cautious control still applies. Indeed, if the same three intermediate reference models above are used, then the intermediate cost functions behave as in Fig. 6.2. Again, at each intermediate step, the global optimum belongs to a candidate DOA of the next one.

6.4 Manipulation of the Reference Spectrum

Candidate domains of attraction have been established by checking, for all ρ in a set and for all $\omega \in [-\pi; \pi]$, the following inequality

$$-\frac{\pi}{2} < \angle S(e^{j\omega}, \rho) - \angle S(e^{j\omega}, \rho_d) < \frac{\pi}{2}. \tag{6.1}$$

This is because under this condition the following matrix is positive-definite

$$M_s(\rho) = \frac{1}{\pi} \int_{-\pi}^{\pi} \Phi_r |G|^2 |S(\rho)|^2 \Re\{S_d^* S(\rho)\} \Re\{\bar{C}\bar{C}^*\} d\omega. \tag{6.2}$$

However, the SPR condition (6.1) is not a necessary condition; it can be circumvented by a proper manipulation of the reference $r(t)$. Even if (6.1) is not satisfied for all ω, the integral in (6.2) will still result in a positive definite matrix $M_s(\rho)$ provided that the reference spectrum is concentrated at those frequencies for which (6.1) is satisfied. But it is not immediately clear how to take advantage of this fact, because this frequency range depends on the unknown process and other unknown quantities, so it is unknown, and at first sight it appears that it might even be empty. Fortunately, there are general structural properties such that this interval is never empty and that allow the designer to concentrate the reference's spectrum within this range. These will be seen next.

6.4.1 Properties of the Sensitivity

For each value of the parameter ρ the sensitivity function is given by:

$$S(z, \rho) = (1 + \rho^T \bar{C}(z) G(z))^{-1}$$

and its phase has the following property for all stabilizing values of ρ.

Lemma 6.1 *For all $\rho^1, \rho^2 \in \Gamma$, the following equalities are satisfied:*

$$\angle S(1, \rho^1) - \angle S(1, \rho^2) = 0$$
$$\angle S(-1, \rho^1) - \angle S(-1, \rho^2) = 0.$$

Proof Let b_i be the poles of the loop transfer function $C(z, \rho)G(z)$ and $\mathrm{a}_i(\rho)$ be the closed-loop poles—that is, the poles of $S(z, \rho)$. Clearly, the b_i are the same for all ρ, because the parameter ρ only enters in the numerator of $C(z, \rho)G(z)$, and these b_i are also the zeros of $S(z, \rho)$. For any frequency ω and any parameter value ρ we can write:

$$\angle S(e^{j\omega}, \rho) = \sum_{i=1}^{n} \angle(e^{j\omega} - \mathrm{b}_i) - \sum_{i=1}^{n} \angle(e^{j\omega} - \mathrm{a}_i(\rho)). \tag{6.3}$$

For $\omega = 0$ (6.3) becomes

$$\angle S(1, \rho) = \sum_{i=1}^{n} \angle(1 - \mathrm{b}_i) - \sum_{i=1}^{n} \angle(1 - \mathrm{a}_i(\rho)). \tag{6.4}$$

Because $\rho \in \Gamma$, for all real poles—that is, $\mathrm{a}_i(\rho) \in \mathbb{R}$, $\angle(1 - \mathrm{a}_i(\rho)) = 0$—see Fig. 6.3a. Also, all complex poles appear in conjugate pairs, and it is easy to see that $\angle(1 - \mathrm{a}_i(\rho)) + \angle(1 - \mathrm{a}_i^*(\rho)) = 0$—see Fig. 6.3b. As a consequence, the second sum

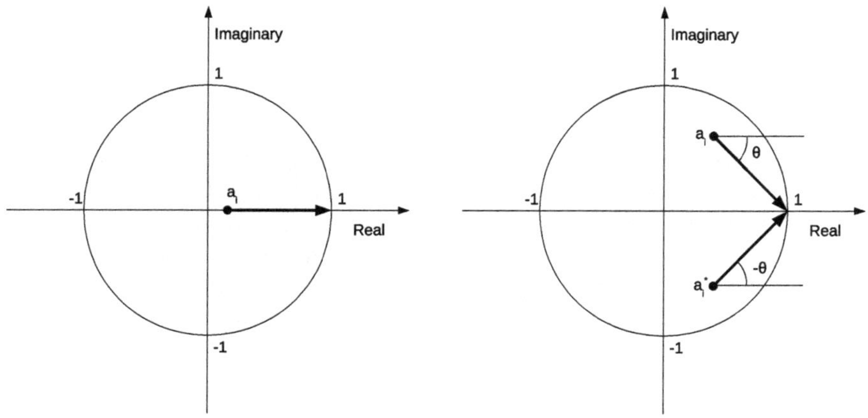

(a) For real poles, $\angle(1 - \mathrm{a}_i(\rho)) = 0$.

(b) For complex conjugate poles, $\angle(1 - \mathrm{a}_i(\rho)) = \angle(1 - \mathrm{a}_i^*(\rho))$.

Fig. 6.3 Determining each term of the phase $\angle S(1, \rho)$

in (6.4) is zero:

$$\sum_{i=1}^{n} \angle(1 - \mathrm{a}_i(\rho)) = 0 \quad \forall \rho \in \Gamma.$$

Then $\angle S(1, \rho) = \sum_{i=1}^{n} \angle(e^{\jmath\omega} - \mathrm{b}_i)$ does not depend on ρ, proving the first inequality in the theorem's statement.

For $\omega = \pi$, $\angle(e^{\jmath\omega} - \mathrm{a}_i(\rho)) = \angle(-1 - \mathrm{a}_i(\rho)) = \pi$ and as a consequence, the second sum in (6.4) is either zero or π, depending on the parity of the system's order:

$$\sum_{i=1}^{n} \angle(-1 - \mathrm{a}_i(\rho)) = \pi \quad \forall \rho \in \Gamma,\ n \text{ odd}$$
$$\sum_{i=1}^{n} \angle(-1 - \mathrm{a}_i(\rho)) = 0 \quad \forall \rho \in \Gamma,\ n \text{ even}.$$

In either case, this sum does not depend on ρ, which proves the second inequality in the statement. □

From the property above and the continuity of $S(e^{\jmath\omega}, \rho)$ we can also conclude that the phase difference between two sensitivity functions is small for frequencies close to $\omega = 0$ and $\omega = \pi$.

Lemma 6.2 *For all $\rho \in \Gamma$, $\exists \omega_l, \omega_h \in (0, \pi)$ such that*:

$$|\angle S(e^{\jmath\omega}, \rho) - \angle S_d(e^{\jmath\omega})| < \frac{\pi}{2} \quad \forall \omega \le \omega_l$$
$$|\angle S(e^{\jmath\omega}, \rho) - \angle S_d(e^{\jmath\omega})| < \frac{\pi}{2} \quad \forall \omega \ge \omega_h.$$

The phase difference between $S_d(z)$ and $S(z,\rho)$ is small at least at the "ends" of the frequency spectrum—for very small frequencies (close to $\omega = 0$) and very large frequencies (close to $\omega = \pi$). Hence, if the spectrum of the reference is concentrated at the two ends of the frequency spectrum then the matrix $M_s(\rho)$ will be positive definite for all ρ and there will be no other extrema except the global optimum ρ_d. This is formalized in the following theorem.

Theorem 6.1 *Let Assumptions* B_y, LP *and* LI *be satisfied, and let* $r(t)$ *be SRp. Then* $\exists\omega_l, \omega_h \in (0,\pi)$ *such that* $\Phi_r = 0\ \forall\omega \in (\omega_l, \omega_h)$ *implies that* ρ_d *is the unique extremum of* $J_y(\rho)$ *in* Γ.

This theorem is a quite general result: it indicates that global convergence to ρ_d can *always* be achieved by a proper choice of the reference spectrum. Of course, the values of ω_l and ω_h are unknown, but this theorem can nevertheless be used to choose a good reference signal. For example, if the controller has only one or two tuning parameters, it suffices to apply a reference whose spectrum is exactly at the borders of the frequency spectrum, as formalized in the following corollary.

Corollary 6.1 *Let Assumptions* B_y, LP *and* LI *be satisfied. Let* α_1 *and* α_2 *be two nonzero but otherwise arbitrary real numbers. The global minimum* ρ_d *is the unique extremum of* $J_y(\rho)$ *in* Γ

- *for* $p = 1$ *if* $C(-1,\rho)G(-1)$ *is finite for all* $\rho \in \Gamma$ *and* $r(t) = \alpha_1(-1)^t$;
- *for* $p = 2$ *if* $C(-1,\rho)G(-1)$ *and* $C(1,\rho)G(1)$ *are finite for all* $\rho \in \Gamma$ *and* $r(t) = \alpha_1 + \alpha_2(-1)^t$.

It is also important to remember that in picking her/his own reference the designer is actually changing the performance criterion. This is something she/he does not always have the luxury of doing. We will see in the following some different ways to choose the appropriate experimental conditions in order to profit from these properties to obtain global convergence and still reach the desired minimum—the minimum of the original performance criterion with the reference that the practical process actually has to follow.

Before moving into that, let us note that a similar theoretical result can be established for $J_e(\rho)$, though this is of limited practical use because the spectrum Φ_ν cannot be manipulated.

Lemma 6.3 *Let Assumptions* B_e, LP *and* LI *be satisfied. Then* $\exists\omega_l^\nu, \omega_h^\nu$ *such that* $\Phi_\nu = 0\ \forall\omega \in (\omega_l^\nu, \omega_h^\nu)$ *implies that* ρ_e *is the unique extremum of* $J_e(\rho)$ *in* Γ.

6.4.2 Applying a Different Reference

In any control system, the reference that the system is supposed to track is given and is not something the designer can choose. But the designer can still apply a different reference during some noncritical phases of process operation, during which the system can take the opportunity to find out the best controller parameters. Or the designer can add a small signal on top of the reference to be tracked, thus guaranteeing persistence of excitation and other properties necessary for a good adaptation without compromising the tracking performance. In short, there are certainly important constraints to when and how to apply the reference signal that will provide the desired convergence features for the data-driven controller design, but it is nevertheless possible to do so.

Theorem 6.1 states that there always exists a reference for which there are not local maxima or minima to prevent convergence in the entire stability set Γ. Moreover, under the standing Assumption B_y, the global optimum of $J_y(\rho)$ does not depend on the reference spectrum; it is always equal to ρ_d. So, whatever reference the system is supposed to track, it is always possible to choose another reference for which the optimization can be performed from any initial stabilizing controller and that will yield the same final result. In other words, if our cost function is difficult to minimize, we minimize instead another one, which is easier to minimize and has the same global minimum. Minimizing a different cost function with the same minimum is also the central idea in VRFT and some of the fundamental methods in MRAC, an idea that invariably relies on Assumption B_y.

The same argument holds for the noise rejection cost $J_e(\rho)$, but not for the combined cost $J_T(\rho) = J_y(\rho) + J_e(\rho)$. Nevertheless, when minimizing the combined cost $J_T(\rho)$ we can expect $J_e(\rho)$ to be significantly smaller than $J_y(\rho)$ away from ρ_d, in which case the argument holds approximately and the approach will still be effective.

But one should not rely so much on Assumption B_y here. For two significantly different references the global minima can be significantly different, even if Assumption B_y is "almost" satisfied (that is, if $\|K(z)\|$ is small). Changing the cost function is reasonable to eliminate local minima and to calculate the step sizes and so on, but we should still minimize the original cost function. A safer and thus more advisable posture, which does not depend on the exact satisfaction of Assumption B_y to yield the correct results, is to use different references as intermediate steps, similarly to what has been done in cautious control. After all, if the designer is really confident that Assumption B_y is satisfied up to a negligible error, then he/she could resort to VRFT.

The following example illustrates the application of different references to the minimization.

Example 6.2 Consider again the process, the controller structure and the cost function $J_T(\rho)$ in Example 6.1, resulting from the tracking of a reference that is a square wave of unitary amplitude and period $T = 6$ samples with noise variance $\sigma_e^2 = 0.2$. Recall that the corresponding cost function $J_T(\rho)$ possesses a maximum at $\rho \approx 0.6$. In order to get rid of this extremum that prevents convergence to the global minimum from initial controllers with very low gain, let us use the tracking of square-waves with larger periods as intermediate objective functions. In making the period larger, the reference spectrum becomes more concentrated at low frequencies, as can be observed in Fig. 6.4. The result of optimizing the cost $J_T(\rho)$ successively for three

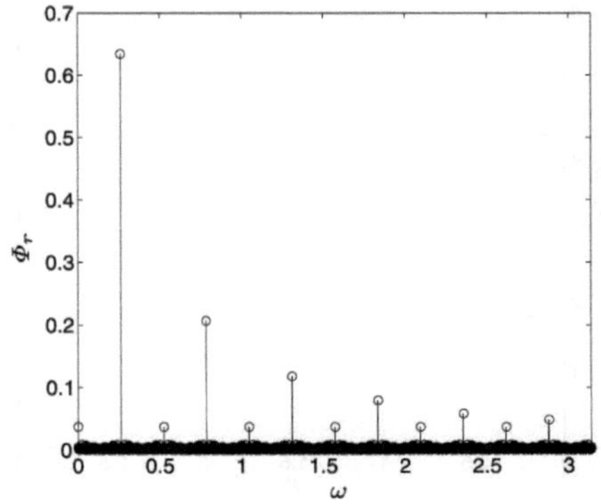

(a) $\Phi_r(e^{\jmath\omega})$ for $T = 24$ samples

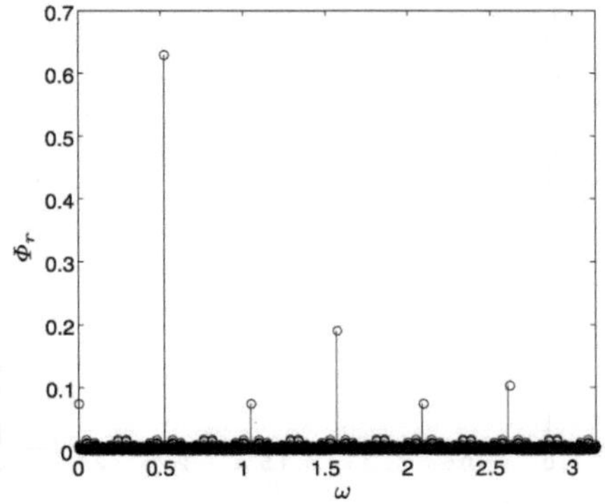

(b) $\Phi_r(e^{\jmath\omega})$ for $T = 12$ samples

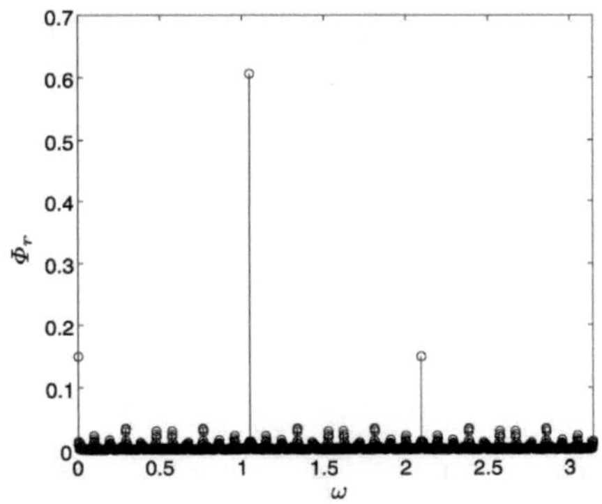

(c) $\Phi_r(e^{\jmath\omega})$ for $T = 6$ samples

Fig. 6.4 Example 6.2—spectra of the square-wave with different periods T

different square waves is shown in Fig. 6.5. It can be observed that for the initial cost, given by the reference with largest period ($T = 24$ samples), the global optimum is at $\rho_*^1 = 1.9$ and Γ is a candidate DOA. The cost with the second reference ($T = 12$ samples) has a minimum around $\rho_*^2 = 2.2$ and there are no other extrema in the set (ρ_*^1, ρ_*^2). Finally, ρ_*^2 is within a candidate DOA for the desired cost function.

Notice in this example how the facts of Theorem 6.1 have been explored. Knowing that for sufficiently low frequencies the maximum would disappear, we tried a reference similar to the original one—a square-wave—only with a lower frequency spectrum. Also, it was not necessary to make the spectrum totally concentrated in the low frequency range, but only to make it much larger in the low frequency range than outside it.

6.4.3 Windowing

Actually applying to the process a reference that is different from the one it is supposed to track is something that, although feasible, may be very inconvenient to do.

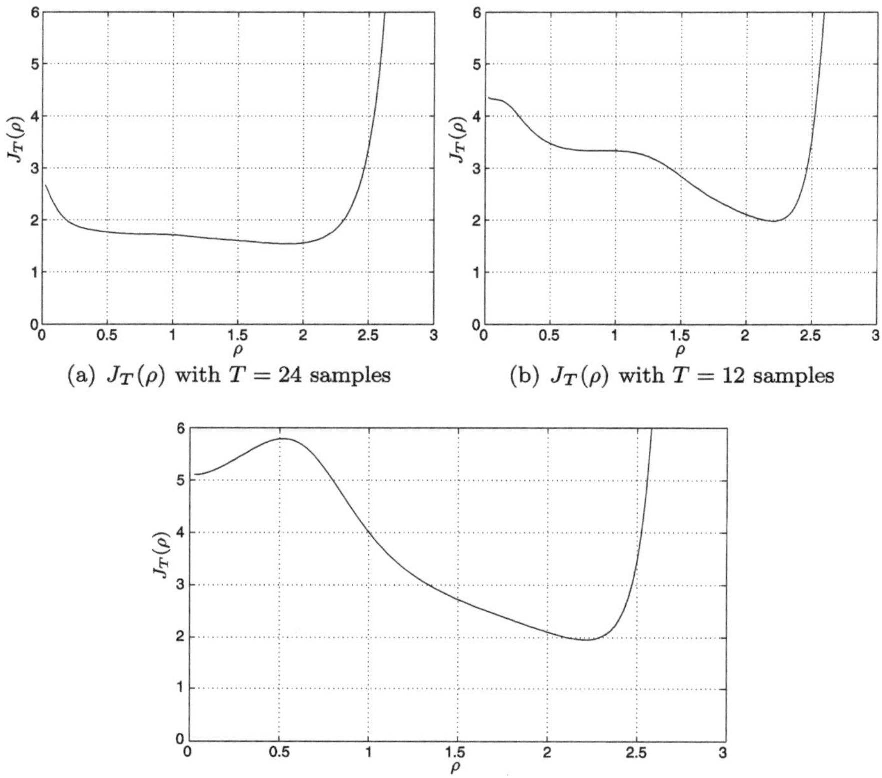

(a) $J_T(\rho)$ with $T = 24$ samples

(b) $J_T(\rho)$ with $T = 12$ samples

(c) $J_T(\rho)$ with $T = 6$ samples

Fig. 6.5 Example 6.2—H_2 performance criterion $J_T(\rho)$ for different reference spectra: square-wave with different periods T

An alternative consists in windowing the data such that the cost function "sees" a different reference spectrum even though the reference is not actually changed. We will not dwell upon the details of windowing methods, which belong to the realm of digital signal processing. We will instead present only a very basic form of windowing to illustrate the central idea and to provide a simple tool that can be applied in many cases for data-driven control design. Still, more advanced windowing techniques are worth exploring.

The basic technique to be presented consists in the use of a rectangular window, whose size and position in the time axis will determine the effective spectrum "seen" by the cost function. We start by noticing that the cost $J_y(\rho)$ as originally defined is not computable in practice, because it involves expectations that can not be calculated, only estimated by a single realization of the stochastic process involved. What can be computed, and is actually computed in real applications, is the following function

$$
\begin{aligned}
\hat{J}_y(\rho, N) &= \frac{1}{N}\sum_{t=1}^{N}(y(t,\rho) - y_d(t))^2 \\
&= \frac{1}{N}\sum_{t=1}^{N}[(T_d(z) - T(z,\rho))r(t) + S(z,\rho)v(t)]^2. \qquad (6.5)
\end{aligned}
$$

If the signal-to-noise ratio is large, then

$$\hat{J}_y(\rho, N) \cong \frac{1}{N}\sum_{t=1}^{N}[(T(z,\rho) - T_d(z))r(t)]^2.$$

Under the standing assumption that all signals are quasi-stationary, the sum above converges to $J_y(\rho)$ as the data window size N grows:

$$\lim_{N\to\infty} \hat{J}_y(\rho, N) = J_y(\rho).$$

We would like to have a reference whose spectrum is concentrated either at very low or at very high frequencies. But the spectrum of the same reference is computed differently when measured with different data sizes. To see this, define the signal $r^N(t)$ obtained as the periodic repetition of a truncation at $t = N$ of $r(t)$, that is

$$r^N(t + kN) \triangleq r(t) \quad t = 1, \ldots, N$$

for all integer k. Now calculate the associated cost $J_y(\rho)$ that would result if this signal were applied to the control system:

$$
\begin{aligned}
J_y(\rho) = \bar{E}[(T(z,\rho) - T_d(z))r^N(t)]^2 &= \lim_{m\to\infty}\frac{1}{m}\sum_{t=1}^{m}[(T(z,\rho) - T_d(z))r^N(t)]^2 \\
&= \lim_{k\to\infty}\frac{1}{kN}k\sum_{t=1}^{N}[(T(z,\rho) - T_d(z))r^N(t)]^2 \\
&= \frac{1}{N}\sum_{t=1}^{N}[(T(z,\rho) - T_d(z))r^N(t)]^2 \\
&= \frac{1}{N}\sum_{t=1}^{N}[(T(z,\rho) - T_d(z))r(t)]^2 \\
&= \hat{J}_y(\rho, N).
\end{aligned}
$$

The approximated cost $\hat{J}_y(\rho, N)$ equals the exact cost $J_y(\rho)$ that would have been obtained should $r^N(t)$ have been applied to the system. Hence, using a different N for the calculations (that is, a window with different size) has the same effect on the cost function as applying a different reference signal. For instance, if the reference is a step, taking smaller N is equivalent to applying a reference with more

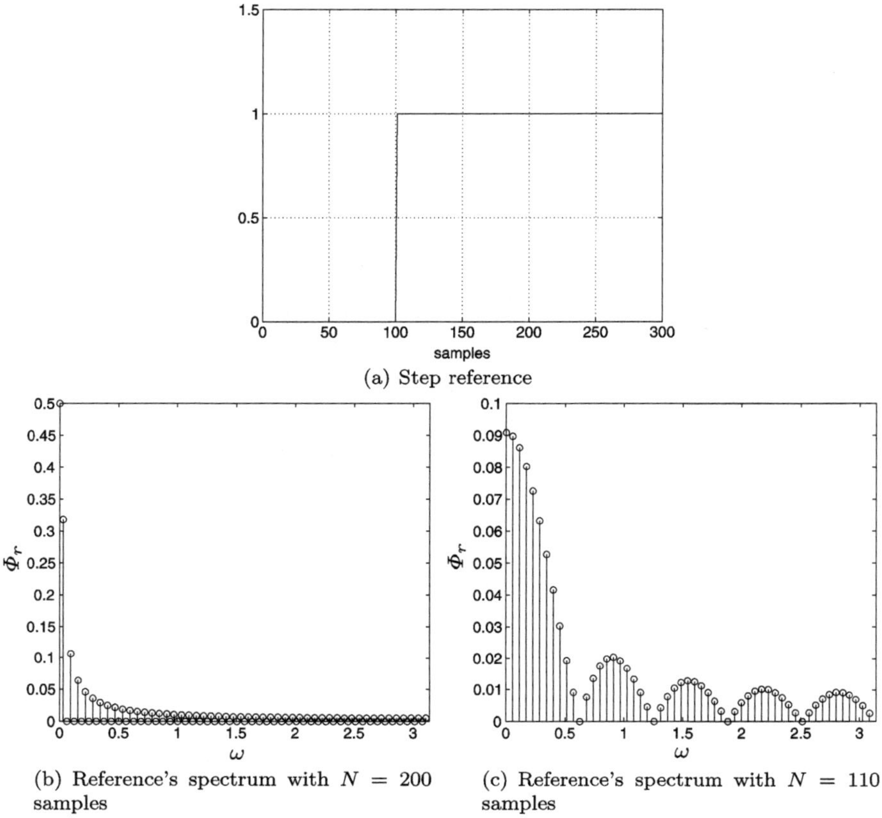

Fig. 6.6 Example 6.3—reference $r(t)$ and its spectrum calculated with different rectangular windows

energy at higher frequencies. On the other hand, taking more data after the transient will make the spectrum more concentrated at low frequencies. This is illustrated by the following example.

Example 6.3 Consider a signal $r(t)$ defined as a unitary step at $t = 100$ samples, that is, $r(t) = 1$ $\forall t > 100$ samples and $r(t) = 0$ $\forall t \leq 100$ samples and let us calculate the spectrum of this signal by the fast Fourier transform. If in this calculation we use $N = 200$ points (that is, a rectangular window covering the first 200 points), the resulting spectrum is the one given in Fig. 6.6b, whereas for $N = 110$ it is given in Fig. 6.6c. Clearly, for this particular class of reference, the spectrum tends to spread out to higher frequencies as we take less data into account. As a measure of this effect, take the amount of the signal energy that is concentrated at each border of the frequency spectrum for each case. For $N = 200$, about 88% of the signal energy is concentrated below $\omega = 0.4\pi$, whereas for $N = 110$ only 77% of the signal energy lies in this range. On the other hand, the range $0.6\pi \leq \omega \leq \pi$ contains about 7% of the signal energy for $N = 200$ and 13% of the signal energy for $N = 110$.

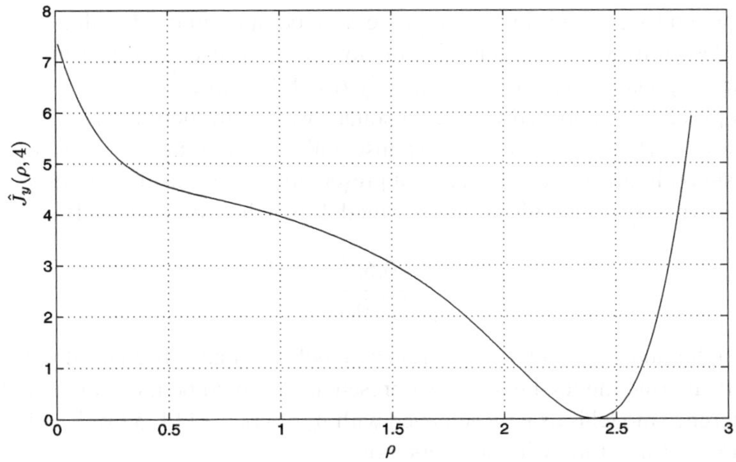

Fig. 6.7 Example 6.4—performance criterion $\hat{J}_y(\rho, 4)$ for square-wave reference, $T = 6$ samples

Thus the cost function can be shaped by changing the size and/or the location of the data window. This is equivalent, regarding its effect on the shape of the cost function, to using a reference with a different spectrum. Choosing an appropriate data window for general reference signals is far from trivial, but it is not so much so for step reference signals, which are the most commonly found in practice. The following example illustrates how to do it.

Example 6.4 Reconsider once again Example 6.2. Take the original reference, with $T = 6$ samples, but include only four data in the cost function, that is, consider the minimization of $\hat{J}_y(\rho, 4)$. This cost function presents no extrema in Γ other than the global minimum ρ_d, as shown in Fig. 6.7.

6.5 Case Studies

6.5.1 A PID Design

To further illustrate the possibilities of cost function shaping in control design, let us consider the design of a PID controller for the following process:

$$G(z) = \frac{1}{100}\frac{z^2 - 0.25}{z^3 - 1.9z^2 + 0.91z}.$$

The transfer function of the PID controller is given by

$$C(z, \rho) = \frac{\varrho_2 z^2 + \varrho_1 z + \varrho_0}{z(z-1)}.$$

Let us pretend we do not know the process model and proceed with a data-driven control design using as the only information on the process the fact that its open-loop step response settles in approximately $t_1 = 80$ samples.

This process is supposed to track a square-wave with period $T_f = 20$ samples. The desired performance is a step response with settling time $t_s \approx 10$ samples with no overshoot. In the interest of clarity of presentation, we consider a reference tracking objective—$J_y(\rho)$—with a reference model such that Assumption B_y is satisfied:

$$T_d(z) = \frac{0.6z^2 - 0.15}{z^3 - 0.4z^2 - 0.15}.$$

A simpler and more intuitive reference model could have been used without significantly altering the final results, but presentation would be far more complicated. This reference model is exactly achieved with $\rho_d = [60 \;\; -114 \;\; 54.6]^T$ and provides the desired settling time with no overshoot.

Since very little knowledge about the process is available, we initially apply a PI controller with conservative tuning, that is with pass-band smaller than that of the process. In this case, the following initial controller qualifies

$$C_0(z) = C(z, \rho_0) = 30\frac{z - 0.9}{z - 1}$$

which corresponds to the initial parameters $\rho_0 = [30 \; -27 \;\; 0]^T$.

If we apply this controller to the process then the settling time t_s and maximum overshoot M_o in response to a step reference are $t_s = 55$ samples and $M_o = 76\%$, respectively. Let us take this PI controller as an initial controller and consider the minimization of $J_y(\rho)$. First let us check the SPR condition (5.17) for this controller

$$\frac{S_d(z)}{S(z, \rho_0)} = \frac{z(z - 0.90)(z - 0.086)(z^2 - 1.61z + 0.87)}{(z - 0.70)(z^2 + 0.30z + 0.21)(z^2 - 1.9z + 0.91)}.$$

The phase of this transfer function is shown in Fig. 6.8. We see that the SPR condition is not satisfied: the phase difference between $S_d(z)$ and $S(z, \rho_0)$ is larger than $90°$ in the interval $0.16 < \omega < 0.45$. On the other hand, the fundamental frequency of the reference is $\omega_f = \frac{2\pi}{20} = 0.316$, which is exactly within this range.

So, it is expected to have troublesome extrema of the cost function around ρ_0. It is not possible to see the cost function's dependence on the parameter vector in this case (because there are three parameters), but it is possible to see its shape along directions in the parameter space. One such direction is presented in Fig. 6.9, specifically the variation of $J_y(\rho)$ along the line that connects the global optimum ρ_d and the initial parameter value ρ_0. In this plot the abscissa is parameterized as $\rho = \rho_d + a(\rho_0 - \rho_d)$; so, $a = 1$ corresponds to ρ_0 and $a = 0$ to ρ_d. It is observed that at $a \cong 0.85$ the curve presents a maximum. This is not a maximum of the cost function, only a constrained maximum along this line, but still it implies that at this point the component of the gradient $\nabla J_y(\rho)$ along this direction is zero—$(\rho - \rho_d)^T \nabla J_y(\rho) = 0$. Moreover, in the interval $0.85 > a > 1.1$, which includes the initial controller, $(\rho - \rho_d)^T \nabla J_y(\rho) < 0$. Hence, no candidate DOA contains ρ_0.

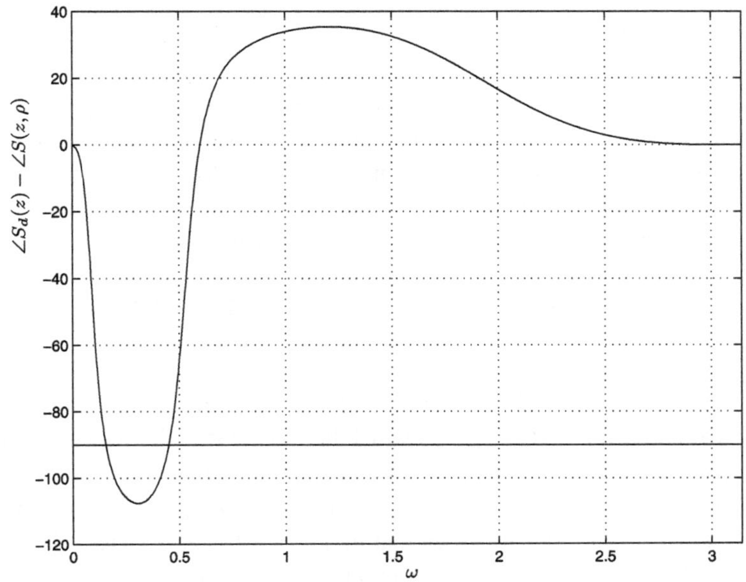

Fig. 6.8 Case study—the distance between the initial and the desired sensitivities

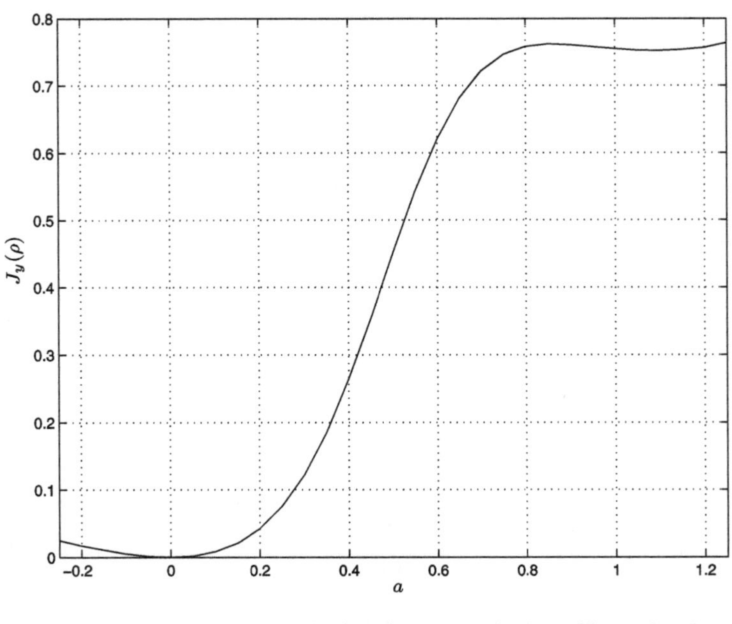

Fig. 6.9 Case study—cost function along the direction $\rho_0 - \rho_d$ for $T_f = 20$ samples; the parameter is normalized: $\rho = \rho_d + a(\rho_0 - \rho_d)$

6.5.1.1 Cautious Control

Trying to optimize the desired performance criterion starting from ρ_0 is very unlikely to result in convergence to the global optimum and may even lead to instability. So, let us approach the design in a more cautious way, using a less ambitious reference model, that is, one whose performance is slightly better than the one obtained with the initial controller. For instance, the reference model

$$T_d^1(z) = \frac{0.45z^4 - 0.71z^3 + 0.16z^2 + 0.18z - 0.068}{z^5 - 2.45z^4 + 2.11z^3 - 0.75z^2 + 0.18z - 0.068}$$

yields a settling time $t_s = 22$ samples and an overshoot $M_o = 29\%$ and is exactly achieved with $\rho_d^1 = [45\ {-70.5}\ 27.3]^T$, which lies on the straight line connecting ρ_d and ρ_0; more specifically, $\rho_d^1 = \rho_d + a(\rho_0 - \rho_d)$ for $a = 0.5$.

Then the intermediate cost function $J_y^1(\rho) \triangleq E[((T(z,\rho) - T_d^1(z))r(t))^2]$ behaves as shown in Fig. 6.10. It is seen that $J_1(\rho)$ vanishes at $a = 0.5$, which corresponds exactly to ρ_d^1, and that there are no other extrema of the cost function in this direction.

Analysis in other directions yield similar results—no extrema are found by inspection—so we can expect to find a DOA for $J_y^1(\rho)$ which contains ρ_0. Define, for example, the polyhedron Υ_1 whose vertices are given by:

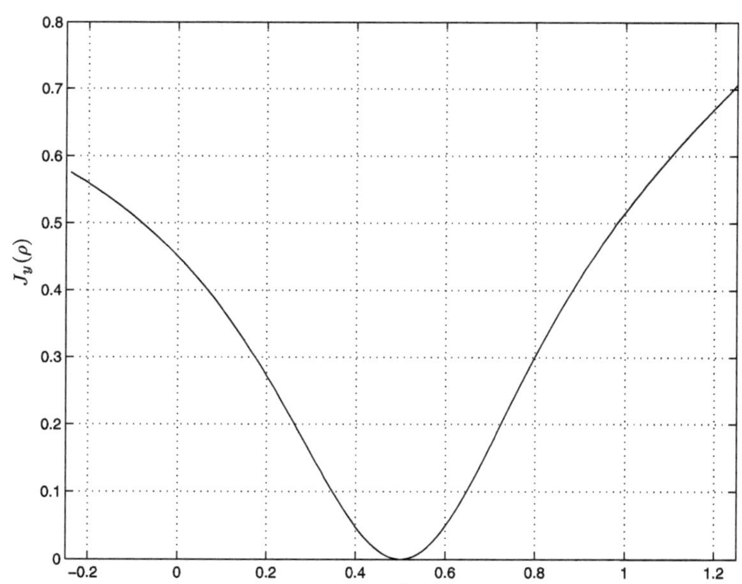

Fig. 6.10 Case study with cautious control—intermediate cost function $J_y^1(\rho)$ along the direction $\rho_0 - \rho_d$ for $T_f = 20$; the parameter is normalized: $\rho = \rho_d + a(\rho_0 - \rho_d)$

$$\begin{bmatrix} 45.50 \\ -73.04 \\ 28.11 \end{bmatrix} \begin{bmatrix} 45.19 \\ -72.88 \\ 29.05 \end{bmatrix} \begin{bmatrix} 46.31 \\ -72.47 \\ 28.28 \end{bmatrix} \begin{bmatrix} 46.00 \\ -72.31 \\ 29.22 \end{bmatrix}$$

$$\begin{bmatrix} 29.00 \\ -25.19 \\ -1.92 \end{bmatrix} \begin{bmatrix} 28.69 \\ -25.03 \\ -0.98 \end{bmatrix} \begin{bmatrix} 29.81 \\ -24.62 \\ -1.75 \end{bmatrix} \begin{bmatrix} 29.50 \\ -24.46 \\ -0.81 \end{bmatrix}.$$

This polyhedron contains ρ_0 and ρ_d^1 and it is easy, although somewhat time consuming and very boring, to define a fine grid within this polyhedron and verify numerically that, for each ρ of this grid, the following condition is satisfied:

$$(\rho - \rho_d^1)^T J_y^1(\rho) > 0. \tag{6.6}$$

Any ball inside Υ_1 is a candidate DOA for the cost $J_y^1(\rho)$ and it is thus possible to converge to ρ_d^1 taking ρ_0 as initial condition in a steepest descent algorithm applied to the cost function $J_y^1(\rho)$.

Once ρ_d^1 is reached, the system's performance has already been significantly improved regarding the initial controller. Indeed the original performance criterion evaluates to $J_y(\rho_d^1) = 0.47$—as can be seen in Fig. 6.9—whereas $J_y(\rho_0) = 0.76$. Yet, this is just an intermediate step.

Now observe in Fig. 6.9 that there are no extrema of $J_y(\rho)$ (the original performance criterion) between ρ_d and ρ_d^1, so we can expect to find a candidate DOA for $J_y(\rho)$ which contains ρ_d^1. Indeed, let Υ be the polyhedron defined by the following vertices

$$\begin{bmatrix} 60.70 \\ -116.27 \\ 55.83 \end{bmatrix} \begin{bmatrix} 60.60 \\ -116.21 \\ 56.05 \end{bmatrix} \begin{bmatrix} 60.90 \\ -116.14 \\ 55.88 \end{bmatrix} \begin{bmatrix} 60.80 \\ -116.08 \\ 56.10 \end{bmatrix}$$

$$\begin{bmatrix} 44.20 \\ -68.42 \\ 25.80 \end{bmatrix} \begin{bmatrix} 44.10 \\ -68.36 \\ 26.02 \end{bmatrix} \begin{bmatrix} 44.40 \\ -68.29 \\ 25.85 \end{bmatrix} \begin{bmatrix} 44.30 \\ -68.23 \\ 26.07 \end{bmatrix}.$$

Observe that Υ contains ρ_d^1 and ρ_d. By defining a fine grid within Υ and verifying numerically that, for each ρ of this grid, the following condition is satisfied:

$$(\rho - \rho_d)^T J_y(\rho) > 0 \tag{6.7}$$

it is established that any ball inside Υ is a candidate DOA for $J_y(\rho)$ and it is thus possible to converge from ρ_d^1 to ρ_d by means of a steepest descent algorithm applied to the cost function $J_y(\rho)$.

Thus a two step cautious control approach allows to obtain convergence, with a steepest descent optimization, from the poor initial condition ρ_0 to the global minimum ρ_d. Notice that the knowledge of the process' transfer function was used along the example to verify that this was indeed possible, but not to decide how to use the intermediate reference model $T_d^1(z)$.

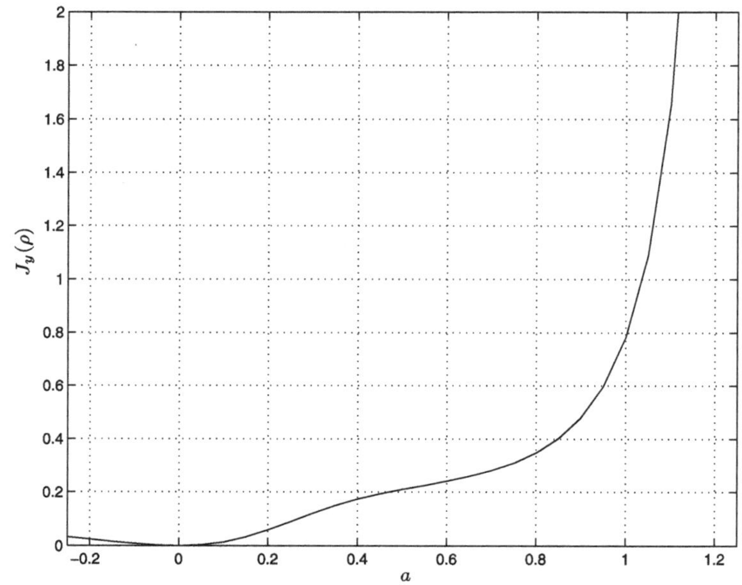

Fig. 6.11 Case study—cost function along the direction $\rho_0 - \rho_d$ for $T_f = 30$; the parameter is normalized: $\rho = \rho_d + a(\rho_0 - \rho_d)$

6.5.1.2 Shaping the Reference Spectrum

For the reference to be tracked—a square wave with period 20 samples—convergence to the global optimum is not obtained from the initial controller given. Let us consider changing the reference spectrum to properly shape the cost function. This can be done in two ways: either applying a different reference or changing the size of the data window.

By increasing the period of the reference to $T'_f = 30$ samples, the cost function behaves as shown in Fig. 6.11. It can be seen that there are no extrema along this line for all parameter values in Γ. Just like in the cautious control case, analysis in other directions show similar behavior and candidate DOAs can be found by inspection. Alternatively, we can keep the same reference applied to the system and take less data into account in the calculation of the cost. Recall that this has the opposite effect on the reference spectrum regarding the change in reference proposed above: it tends to concentrate the spectrum in higher frequencies. The resulting cost function behaves as shown in Fig. 6.12.

6.5.2 A Noisy Process

Consider the following system

$$y(t) = \frac{0.15}{z - 0.5} u(t) + e(t),$$

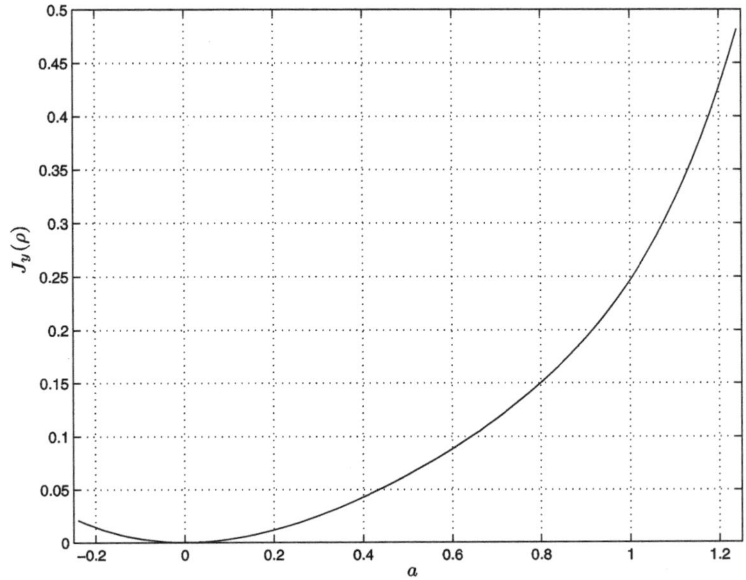

Fig. 6.12 Case study—cost function along the direction $\rho_0 - \rho_d$ for $T_f = 30$ samples; the parameter is normalized: $\rho = \rho_d + a(\rho_0 - \rho_d)$

where $e(t)$ is white noise with variance $\sigma_e^2 = 0.001$. The system is controlled by a PI controller

$$C(z,\rho) = \begin{bmatrix} \varrho_1 & \varrho_2 \end{bmatrix} \begin{bmatrix} \frac{z}{z-1} \\ \frac{1}{z-1} \end{bmatrix}.$$

The desired reference model is

$$T_d = \frac{0.1}{z-0.9}$$

which can be exactly achieved with the ideal (PI) controller

$$C_d(z) = \begin{bmatrix} 2/3 & -1/3 \end{bmatrix} \begin{bmatrix} \frac{z}{z-1} \\ \frac{1}{z-1} \end{bmatrix} \in \mathcal{C}.$$

The following controller is initially running in the loop

$$C(z,\rho_0) = \begin{bmatrix} 2 & 2 \end{bmatrix} \begin{bmatrix} \frac{z}{z-1} \\ \frac{1}{z-1} \end{bmatrix}.$$

Figure 6.13 shows the closed-loop response with this controller and the desired behavior. We can see that the output is very different from the desired one. An iterative data-driven control design with the steepest descent optimization will be used to improve the closed-loop behavior.

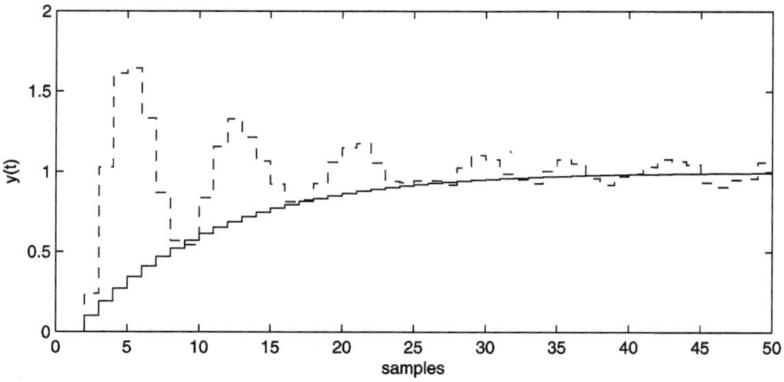

Fig. 6.13 Closed-loop step response with controller $C(z, \rho_0)$ and the desired response

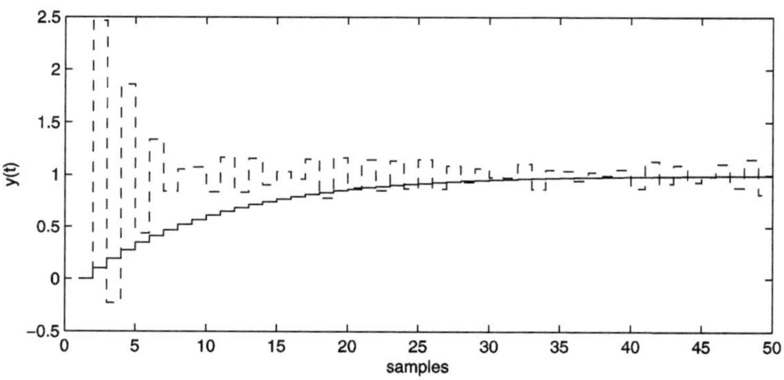

Fig. 6.14 Closed-loop step response with controller $C(z, \rho_{100})$ and the desired response

We choose the following reference signal, which is SR3, to use in the experiments.

$$r_1(t) = 1 + \sin\left(\frac{2\pi t}{10}\right).$$

After 100 iterations the algorithm converged to the following controller:

$$C(z, \rho_{100}) = \begin{bmatrix} 16.2130 & -2.0725 \end{bmatrix} \begin{bmatrix} \frac{z}{z-1} \\ \frac{1}{z-1} \end{bmatrix}.$$

Figure 6.14 shows that the closed-loop response is still far from desired. It is observed in Fig. 6.15 that the cost is being reduced at each iteration, however in Fig. 6.16 we can see that the algorithm is converging to some parameter value distant from $C_d(z)$. Because the system is affected by noise, the algorithm will not converge exactly to $C_d(z)$, but because the noise level is low, the algorithm should converge to some controller close to $C_d(z)$. In fact, the algorithm has converged to a local minimum and realizing this only requires knowledge of the fact that Assump-

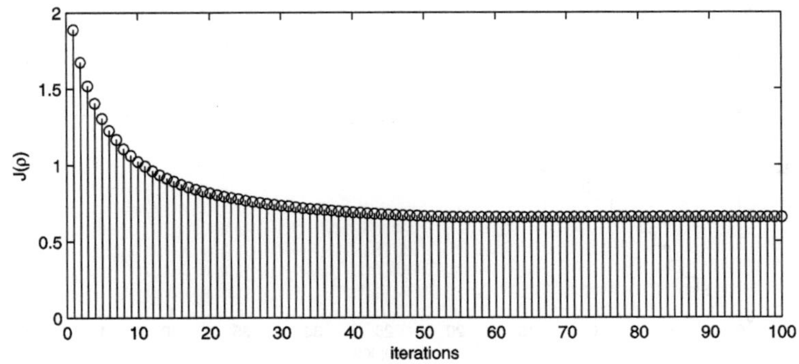

Fig. 6.15 Cost at each iteration—$r_1(t)$

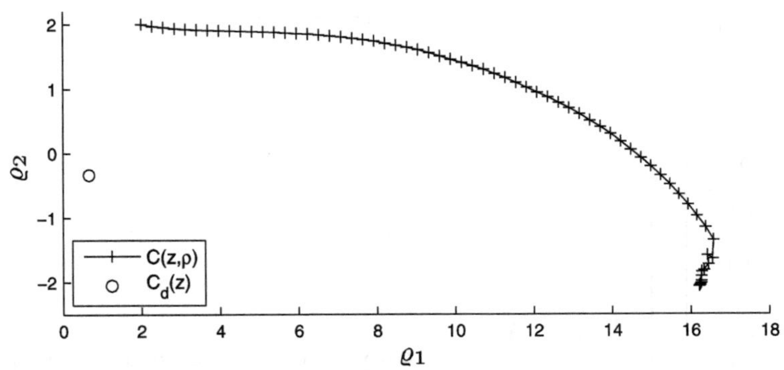

Fig. 6.16 Controller parameters at each iteration—$r_1(t)$—Initial condition $\rho_0 = [2\ 2]^T$

tion B_y is satisfied. Under this assumption, the cost function at the global minimum evaluates at a very low value—of the same order of the noise's variance—and yet the algorithm has lead to values of the cost function which are orders of magnitude larger.

Now let us use another reference signal

$$r_2(t) = 1 + \sin\left(\frac{2\pi t}{5}\right).$$

This signal results in a different cost function, for which the local minimum, at which the optimization has been trapped before, does not exist anymore. Indeed, after 100 iterations the algorithm converged to the controller

$$C(z, \rho_{100}) = \begin{bmatrix} 0.6692 & -0.3398 \end{bmatrix} \begin{bmatrix} \frac{z}{z-1} \\ \frac{1}{z-1} \end{bmatrix},$$

which is very close to $C_d(z)$. The closed-loop response is shown in Fig. 6.17, where it is observed that the performance is close to the desired one. The evolution of the

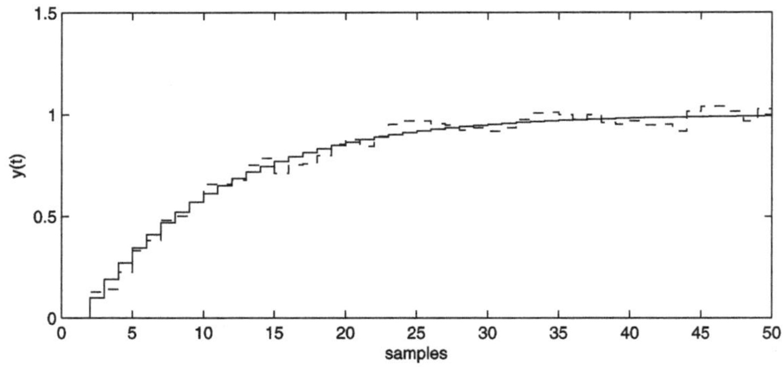

Fig. 6.17 Closed-loop step response with controller $C(z, \rho_{100})$ and the desired response

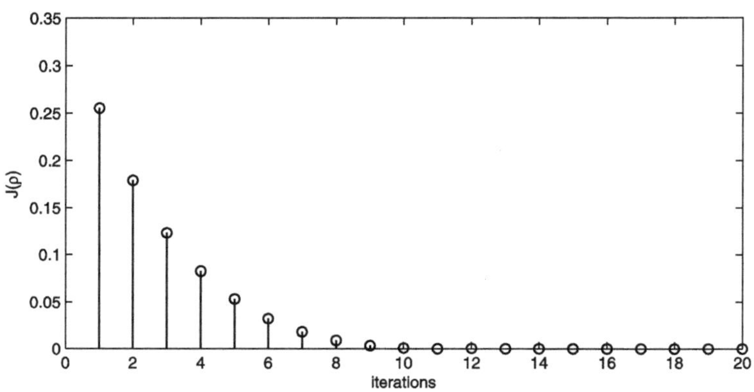

Fig. 6.18 Cost at each iteration—$r_2(t)$

controller's parameter at each iteration can be seen in Fig. 6.19 and the cost at each iteration is observed in Fig. 6.18.

In Fig. 6.20 we can see that $|\angle S_d(e^{j2\pi/5})/S(e^{j2\pi/5}, \rho_0)| < 90°$ while $|\angle S_d(e^{j2\pi/10})/S(e^{j2\pi/10}, \rho_0)| > 90°$.

6.6 Chapter Conclusions

Cost function shaping makes use of the variables that the designer has at his/her disposal: the reference signal, the intermediate reference models, the data window and the initial controller. Regarding the reference signal, applying references with frequency content sufficiently constrained to low and/or high frequencies is appropriate. Different reference spectra can also be simulated by taking different data windows with the same applied reference. For a step reference, for instance, it is clear how to do this: taking very few data right after the step concentrates the apparent spectrum in high frequencies, whereas taking very large sets of data has the

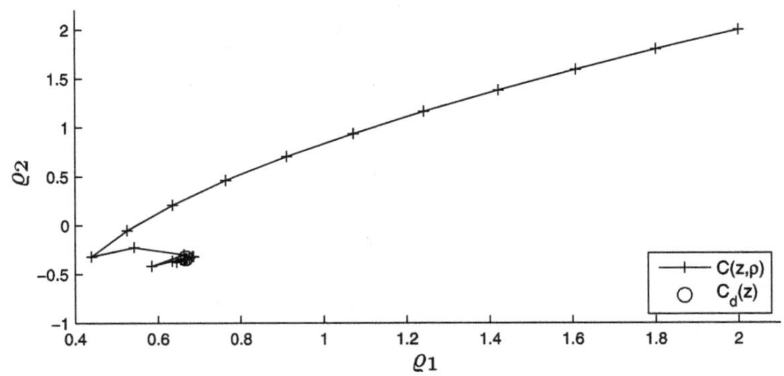

Fig. 6.19 Controller parameters at each iteration—$r_2(t)$—Initial condition $\rho_0 = [2\ 2]^T$

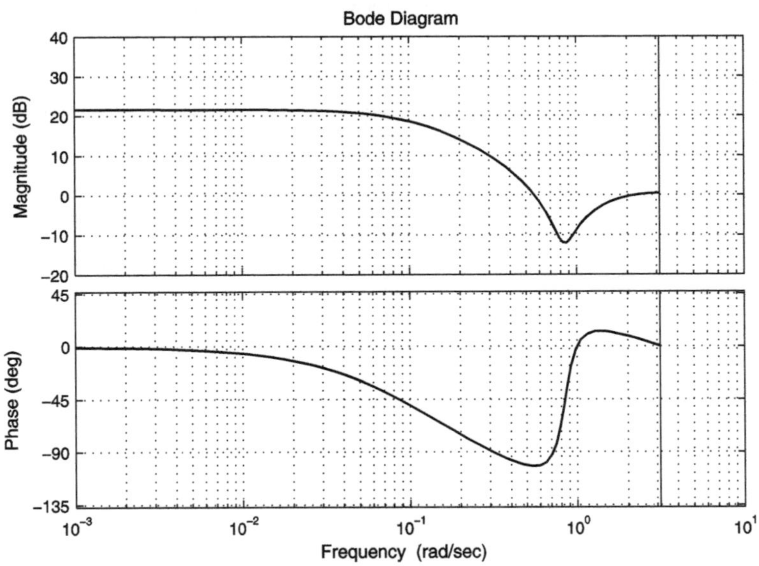

Fig. 6.20 Bode diagram of $S_d(z)/S(z, \rho_0)$

opposite effect. The idea of restricting the reference spectrum to obtain global convergence has appeared in the literature a long time ago [6], but it was not much explored until recently [1].

Concerning the intermediate reference models, they should be sufficiently close to each other so that the global minimum of one intermediate criterion is within a candidate DOA of the next, but not too close so that too many intermediate steps are required to reach the desired performance—which would require too much time. This concept of cautious control is a familiar one in data-driven control design [2, 3] and in iterative identification and control design [7, 8]; it is also a fundamental precept of the windsurfer approach to adaptive control [4, 5]. In short it consists of the route pursued in the examples presented in this chapter: modest, easily achievable performance is initially required, and increasingly tight performance requirements are imposed.

References

1. A.S. Bazanella, M. Gevers, L. Mišković, B.D.O. Anderson, Iterative minimization of H_2 control performance criteria. Automatica **44**(10), 2549–2559 (2008)
2. H. Hjalmarsson, S. Gunnarsson, M. Gevers, A convergent iterative restricted complexity control design scheme, in *33rd IEEE Conference on Decision and Control*, Lake Buena Vista, USA, 1994
3. L.C. Kammer, Stability assessment for cautious iterative controller tuning. Automatica **41**, 1829–1834 (2005)
4. W.S. Lee, B.D.O. Anderson, R. Kosut, I.M.Y. Mareels, On robust performance improvement through the windsurfer approach to adaptive robust control, in *32nd IEEE Conference on Decision and Control*, 1993
5. W.S. Lee, B.D.O. Anderson, I.M.Y. Mareels, R.L. Kosut, On some key issues in the windsurfer approach to adaptive robust control. Automatica **31**(11), 1619–1636 (1995)
6. B. Riedle, L. Praly, P.V. Kokotovic, Examination of the SPR condition in output error parameter estimation. Automatica **22**(4), 495–498 (1986)
7. E. Trulsson, L. Ljung, Adaptive control based on explicit criterion minimization. Automatica **21**, 385–399 (1985)
8. Z. Zang, R.R. Bitmead, M. Gevers, Iterative weighted least-squares identification and weighted LQG control design. Automatica **31**(11), 1577–1594 (1995)

Chapter 7
Computations

In the previous chapters we have given a general formulation, and a general theory, for the iterative optimization of H_2 control performance criteria. Actually implementing this optimization with the classical methods (steepest descent, Newton-Raphson and their variants) is far from trivial. In order to do that it is necessary to compute, at each iteration step, at least the gradient of the function being minimized. This gradient is a fairly complicated function of the plant dynamics and of the noise spectrum. When these are unknown, it is not clear how this gradient can be computed.

Yet, the ingenuity of our fellow researchers has provided several procedures to estimate the gradient and other quantities related to the objective function that are required at each step of the optimization. Each one of these procedures was born from a particular perspective taken at the problem at hand and, as such, each one results in a different algorithm that can be applied to a different range of systems and performance criteria. Most of them have been baptized by their authors according to the method's most appealing characteristic, yet some have remained unnamed.

We will see in the following the description of three different schemes for estimation of the gradient and the Hessian in different situations. The methods to be presented are the following: Iterative Feedback Tuning (IFT), Frequency Domain Tuning (FDT) and Correlation based Tuning (CbT). Each one of these methods is described in a separate section of this chapter, and they are organized in chronological order. IFT is oldest (1994), whereas the other methods are more recent and invariably refer to this original IFT paper. So, IFT comes first, followed by FDT (2000) and CbT (2003). Other methods and variants exist, of course, but unfortunately we can not cover all of them in one book. We have selected a range of methods which illustrate different approaches that can be taken to the problem at hand. Indeed, each one of these three methods optimizes a different H_2 performance criterion, and each one derives its gradient/Hessian estimation mechanism from a conceptually different viewpoint.

The FDT method is based on spectral analysis of the signals involved in the normal closed-loop operation and aims at minimizing the noise rejection criterion $J_e(\rho)$. All the analysis and the derivations are made from nonparametric models for

A. Sanfelice Bazanella et al., *Data-Driven Controller Design*,
Communications and Control Engineering,
DOI 10.1007/978-94-007-2300-9_7,

the plant and the controller. The IFT algorithm is derived directly from the time domain analysis of the signals involved in the composite performance criterion $J_T(\rho)$. The CbT method, on the other hand, searches for the minimum of the reference tracking criterion $J_y(\rho)$, and it does so indirectly, by using an instrumental variable approach.

7.1 Iterative Feedback Tuning

Iterative Feedback Tuning (IFT) was proposed by Håkan Hjalmarsson, Svante Gunnarsson and Michel Gevers in the 1994 IEEE CDC [6], whereas the other methods to be presented are more recent and consistently refer to this original paper. This paper is often seen as the landmark for the dawn of data-driven control design. There were previous publications advocating the model-free tuning of controller parameters in much the same philosophy, as far back as in the 1960s, but it was only starting with this paper in 1994 that data-driven design started acquiring a personality of its own. The name Iterative Feedback Tuning (IFT) was coined later on, in a subsequent journal paper of the same authors [5]. Later developments on IFT extended the original algorithm to treat broader classes of systems and performance criteria, and also reported practical applications, including to nonlinear and multivariable systems [1, 3, 4, 11, 12, 14].

We will see in the following the description of the procedure and its implications. The presentation of the other methods, which will be seen in the following sections, follows the same pattern, *mutatis mutandis*.

7.1.1 Derivation

The IFT methodology consists in a set of maneuvers to obtain an unbiased estimate of the gradient and the Hessian for optimization of the composite cost $J_T(\rho)$. Actually the original derivation of IFT was done for an objective function with a penalty in the control effort, something that we did not consider along this book for the reasons discussed at its early chapters. Yet, because considering this penalty in the control effort does not implies any extra burden on the reader at this point, and in order to respect the IFT tradition, we will consider the following H_2 objective function, where $\lambda \in \mathbb{R}^+$ is a design parameter

$$J_{IFT}(\rho) = \bar{E}[(y_d(t) - y(t,\rho))^2 + \lambda u^2(t,\rho)] = J_T(\rho) + \lambda \bar{E}[u^2(t,\rho)]. \tag{7.1}$$

Taking the derivative of (7.1) with respect to the parameter ρ yields

$$\frac{\partial J_{IFT}(\rho)}{\partial \rho} = 2\bar{E}\left[[y_d(t) - y(t,\rho)]\frac{\partial y(t,\rho)}{\partial \rho} + \lambda u(t,\rho)\frac{\partial u(t,\rho)}{\partial \rho} \right].$$

In order to implement the iterative optimization, it is necessary to compute this gradient. This is not possible to compute exactly because it involves expectations of stochastic processes, which are unknown. What can be done is to estimate this expectation by a given single realization of the stochastic process, that is:

$$\frac{\partial \widehat{J_{IFT}(\rho)}}{\partial \rho} = \frac{2}{N}\sum_{t=1}^{N}[y_d(t) - y(t,\rho)]\frac{\partial y(t,\rho)}{\partial \rho} + \lambda \frac{2}{N}\sum_{t=1}^{N} u(t,\rho)\frac{\partial u(t,\rho)}{\partial \rho} \tag{7.2}$$

where N is the number of data collected. Then, with this estimate, the optimization problem can be solved by using a stochastic approximation, that is, replacing the gradient by its estimate. If this estimate turns out to be an unbiased one, then convergence is assured in terms of the Stochastic Approximation Theorem 4.5 and of the robust convergence Theorem 4.6.

Let us work this estimate out in such a way that it will be unbiased. The estimate in (7.2) contains directly measured quantities—$y_d(t)$, $y(t,\rho)$ and $u(t,\rho)$—and the partial derivatives $\frac{\partial y(t,\rho)}{\partial \rho}$ and $\frac{\partial u(t,\rho)}{\partial \rho}$, which are not measured. In order to calculate the gradient in (7.2) it is necessary to somehow synthesize these partial derivatives. Let us start with the partial derivative of the output with respect to the parameter $\frac{\partial y(t,\rho)}{\partial \rho}$. To calculate it, recall that

$$y(t,\rho) = T(z,\rho)r(t) + S(z,\rho)v(t)$$
$$T(z,\rho) = \frac{C(z,\rho)G(z)}{1 + C(z,\rho)G(z)}$$
$$S(z,\rho) = \frac{1}{1 + C(z,\rho)G(z)}.$$

Then we have

$$\frac{\partial y(t,\rho)}{\partial \rho} = \frac{G(z)}{1 + C(z,\rho)G(z)}\frac{\partial C(z,\rho)}{\partial \rho}r(t) - \frac{C(z,\rho)G^2(z)}{(1 + C(z,\rho)G(z))^2}\frac{\partial C(z,\rho)}{\partial \rho}r(t) - \frac{G(z)}{(1 + C(z,\rho)G(z))^2}\frac{\partial C(z,\rho)}{\partial \rho}v(t)$$

which can be rewritten as:

$$\frac{\partial y(t,\rho)}{\partial \rho} = \frac{1}{C(z,\rho)}\frac{\partial C(z,\rho)}{\partial \rho}\{T(z,\rho)r(t) - T(z,\rho)[T(z,\rho)r(t) + S(z,\rho)v(t)]\}. \tag{7.3}$$

This expression for the derivative $\frac{\partial y(t,\rho)}{\partial \rho}$ consists of the application of the vector filter

$$\frac{1}{C(z,\rho)}\frac{\partial C(z,\rho)}{\partial \rho} \triangleq Q(z,\rho) \tag{7.4}$$

to the signal within braces

$$T(z,\rho)r(t) - T(z,\rho)[T(z,\rho)r(t) + S(z,\rho)v(t)] \triangleq \vartheta(t). \tag{7.5}$$

The derivative can be computed if we know the signal $\vartheta(t)$ in (7.5) and then apply the filter (7.4) to it. The filter (7.4) consists of a known transfer function, since it depends only on the controller and its parameterization; no problem there. The signal $\vartheta(t)$ to which the vector filter $Q(z,\rho)$ must be applied, however, is neither a measured signal nor can it be directly calculated from measured quantities, because it contains the unknown transfer functions $T(z,\rho)$ and $S(z,\rho)$ and the noise signal $\nu(t)$. Synthesizing this signal from direct measurements alone, without any process model or prior information, becomes the key ingredient of the IFT development.

Let us regroup the expression of this signal as

$$\vartheta(t) = T(z,\rho)\{r(t) - [T(z,\rho)r(t) + S(z,\rho)\nu(t)]\}.$$

Then, since $y(t,\rho) = T(z,\rho)r(t) + S(z,\rho)\nu(t)$, it can be further regrouped as

$$\vartheta(t) = T(z,\rho)\{r(t) - y(t,\rho)\}. \tag{7.6}$$

Now imagine the following special experiment. There is no noise ($\nu(t) \equiv 0$) in the system and the closed-loop system is excited with the reference $r(t) - y(t,\rho)$, that is, the difference between the reference and the output signals observed in another previously performed experiment. Then, according to (7.6), the output of the closed-loop system in this special experiment would be exactly the signal $\vartheta(t)$ that we are looking for. This observation lead to the suggestion of the following procedure.

First collect the output of the process $y(t,\rho)$ under normal operation, where the reference signal is some given reference $r(t)$. Next, excite the closed-loop system with $r(t) - y(t,\rho)$ and collect the corresponding output again; this second output is given by $\vartheta(t) + S(z,\rho)\nu(t)$, that is, it is a noisy version of the desired signal $\vartheta(t)$ and as such will be considered as an estimate of it. It is important to realize that, since $\nu(t)$ has zero mean, this is an unbiased estimate of $\vartheta(t)$, that is:

$$E[\vartheta(t) + S(z,\rho)\nu(t)] = \vartheta(t).$$

Finally, filter this signal through the appropriate filter $Q(z,\rho)$ in (7.3), thus obtaining the partial derivative $\frac{\partial y(t,\rho)}{\partial \rho}$.

A similar treatment can be given to the other partial derivative $\frac{\partial u(t,\rho)}{\partial \rho}$. Deriving the closed-loop relationship

$$u(t,\rho) = \frac{C(z,\rho)}{1 + C(z,\rho)G(z)}[r(t) - \nu(t)] = S(z,\rho)C(z,\rho)[r(t) - \nu(t)]$$

yields

$$\frac{\partial u(t,\rho)}{\partial \rho} = S(z,\rho)\frac{\partial C(z,\rho)}{\partial \rho}[r(t) - \nu(t)] + \frac{\partial S(z,\rho)}{\partial \rho}C(z,\rho)[r(t) - \nu(t)].$$

Using the following relation:

$$\frac{\partial S(z,\rho)}{\partial \rho} = -\frac{1}{C(z,\rho)}\frac{\partial C(z,\rho)}{\partial \rho}T(z,\rho)S(z,\rho)$$

the following manipulations, similar to the ones just applied to the calculation of $\frac{\partial y(t,\rho)}{\partial \rho}$, lead to a convenient expression for the desired derivative

$$\begin{aligned}\frac{\partial u(t,\rho)}{\partial \rho} &= S(z,\rho)\frac{\partial C(z,\rho)}{\partial \rho}[r(t)-v(t)] - \frac{\partial C(z,\rho)}{\partial \rho}T(z,\rho)S(z,\rho)[r(t)-v(t)] \\ &= S(z,\rho)\frac{\partial C(z,\rho)}{\partial \rho}[r(t)-T(z,\rho)r(t)-(1-T(z,\rho))v(t)] \\ &= \frac{\partial C(z,\rho)}{\partial \rho}S(z,\rho)[r(t)-y(t,\rho)] \\ &= Q(z,\rho)\{C(z,\rho)S(z,\rho)[r(t)-y(t,\rho)]\}.\end{aligned}$$

This expression of the partial derivative, just like (7.3), is the filtering of the bracketed signal $C(z,\rho)S(z,\rho)[r(t)-y(t)]$ through the vector filter $Q(z,\rho)$. Similarly to the findings regarding the output derivative, it can be seen that the signal to be filtered is the control signal $u(t,\rho)$ that would be observed in the control loop if the system were excited with a reference $[r(t)-y(t,\rho)]$ and no noise.

In conclusion, both signal derivatives $\frac{\partial y(t,\rho)}{\partial \rho}$ and $\frac{\partial u(t,\rho)}{\partial \rho}$ that form the gradient estimate can be obtained by measuring the controller's output and the process output in an additional experiment, where we feed the system's reference with the output obtained in the first experiment. The IFT procedure then consists of a sequence of measurements and calculations that is detailed below.

The IFT Algorithm At each step i of an iterative data-driven control design, estimate the gradient $\nabla J_{IFT}(\rho_i)$ through the following procedure.

1. collect data from the normal operation of the process; let the reference signal during this measurement period be denoted by $r_1(t)$ and the resulting controller's output and process output respectively by $u_1(t)$ and $y_1(t)$;
2. compute offline the signal $r_1(t) - y_1(t) \triangleq r_2(t)$;
3. perform an additional experiment applying $r_2(t)$ as the reference to the closed-loop system and measure the resulting controller's output and process output; let these measurements be denoted by $u_2(t)$ and $y_2(t)$ respectively;
4. compute

$$\widehat{\frac{\partial y(t,\rho_i)}{\partial \rho}} = \frac{1}{C(z,\rho_i)}\frac{\partial C(z,\rho_i)}{\partial \rho}y_2(t)$$

and

$$\widehat{\frac{\partial u(t,\rho_i)}{\partial \rho}} = \frac{1}{C(z,\rho_i)}\frac{\partial C(z,\rho_i)}{\partial \rho}u_2(t);$$

5. compute

$$\nabla \widehat{J_{IFT}(\rho_i)} = \frac{2}{N}\sum_{t=1}^{N}[y_d(t)-y_1(t)]\widehat{\frac{\partial y(t,\rho_i)}{\partial \rho}} + \lambda\frac{2}{N}\sum_{t=1}^{N}u_1(t)\widehat{\frac{\partial u(t,\rho_i)}{\partial \rho}}.$$

7.1.2 Extensions

7.1.2.1 Other Search Directions

Search directions other than the steepest descent can also be determined using only an estimate of the first-order derivatives provided by IFT. Then the IFT procedure can be equally applied, modifying only the search direction accordingly. To illustrate these extensions of the IFT method, consider for example the following search direction matrices:

$$R_i = \bar{E}\left[\frac{\partial y(t,\rho_i)}{\partial \rho}\frac{\partial y(t,\rho_i)}{\partial \rho}^T + \lambda\frac{\partial u(t,\rho_i)}{\partial \rho}\frac{\partial u(t,\rho_i)}{\partial \rho}^T\right]^{-1}. \tag{7.7}$$

An unbiased estimate of R_i in (7.7) can be obtained simply by using the same computations already used to obtain the gradient's estimate, that is:

$$\hat{R}_i = \frac{1}{N}\sum_{t=1}^{N}\left[\widehat{\frac{\partial y(t,\rho_i)}{\partial \rho}}\widehat{\frac{\partial y(t,\rho_i)}{\partial \rho}}^T + \lambda\widehat{\frac{\partial u(t,\rho_i)}{\partial \rho}}\widehat{\frac{\partial u(t,\rho_i)}{\partial \rho}}^T\right]^{-1} \tag{7.8}$$

where the estimates $\widehat{\frac{\partial y(t,\rho_i)}{\partial \rho}}$ and $\widehat{\frac{\partial u(t,\rho_i)}{\partial \rho}}$ are obtained as described previously.

The inspiration for this particular choice is the Newton-Raphson's method, in which the search direction is defined by the Hessian:

$$R_i = \left(\frac{\partial J^2(\rho)}{\partial \rho^2}\right)^{-1}.$$

Calculating the Hessian for the IFT performance criterion yields, for $\lambda = 0$ (let us consider only this case for simplicity):

$$\frac{\partial J^2(\rho)}{\partial \rho^2} = 2\bar{E}\left[[y_d(t) - y(t,\rho)]\frac{\partial y^2(t,\rho)}{\partial \rho^2} + \frac{\partial y(t,\rho)}{\partial \rho}\frac{\partial y(t,\rho)}{\partial \rho}^T\right]$$

and the search direction (7.7) consists in taking only the second term in this expression, neglecting the first one. As such, the search direction defined in (7.7) is interpreted as an approximation of the Hessian by many, and referred to as an approximate Newton-Raphson.

Of course, an additional search direction alternative is to apply the Newton-Raphson's method or its variable step variation, with a more accurate estimate of the Hessian. The implementation of this alternative requires extra information when compared to the steepest descent and the Gauss-Newton, namely an unbiased estimate of the Hessian. Obtaining such an estimate directly from data, with the purely model-free approach of IFT, is an unsolved problem in general. But for the particular case of zero reference and no weight in the control effort ($\lambda = 0$), the following procedure is available [15].

Let the reference signal be zero, that is, let the data be generated by

$$y(t,\rho) = S(z,\rho)\nu(t) = \frac{1}{1+C(z,\rho)G(z)}\nu(t).$$

Then the expression (7.3) reduces to:

$$\frac{\partial y(t,\rho)}{\partial \rho} = -T(\rho)S(\rho)\frac{\partial C(z,\rho)}{\partial \rho}\nu(t). \tag{7.9}$$

The second derivative in this case is given by

$$\frac{\partial^2 y(t,\rho)}{\partial \rho^2} = -T(\rho)S(\rho)\frac{\partial^2 C(z,\rho)}{\partial \rho^2}\nu(t) + 2T^2(z,\rho)S(z,\rho)\frac{\partial C(z,\rho)}{\partial \rho}\frac{\partial C(z,\rho)}{\partial \rho}^T\nu(t). \tag{7.10}$$

An estimate can be built for this second derivative with the aid of four experiments, in the very same philosophy and analytical treatment used for obtaining the gradient estimates. The first two experiments are the ones described before, that are used to estimate the gradient. The third experiment is a repetition of the second one, and will differ from this one only by the fact that it is driven by a different noise realization. The fourth experiment consists in applying to the reference the output of the third experiment. These additional experiments allow to construct an estimate of the second derivative:

$$\frac{\widehat{\partial^2 y(t,\rho)}}{\partial \rho^2} = 2\frac{\partial C(z,\rho)}{\partial \rho}\frac{\partial C(z,\rho)}{\partial \rho}^T y^4(t,\rho) - \frac{\partial^2 C(z,\rho)}{\partial \rho^2}y^2(t,\rho)$$

where $y^2(t,\rho)$ and $y^4(t,\rho)$ are the outputs of the second and fourth experiment respectively.

In this way, unbiased estimates of the Hessian can be obtained at the cost of two more experiments. The one special experiment necessary to estimate the gradient (the second one) is already a potential nuisance that somehow limits the practical application of IFT, so the realization of yet two more special experiments is a possibility to be considered with prudence.

The approximate Newton-Raphson algorithm, defined by the search direction (7.7), tends to present faster convergence than the steepest descent, without the extra information and experiments required by the Newton-Raphson's algorithm, or its extra risks arriving from its typically smaller domain of attraction. As a result, IFT applications have been most successful with this approximate Newton-Raphson algorithm, even though the appropriate choice of the step sizes for the steepest descent, as presented in Chap. 5, somewhat levels the field between the two algorithms.

7.1.2.2 Two Degree-of-Freedom Controllers

The IFT procedure can also be applied to the case of controllers with two degrees of freedom—2DOF. A 2DOF controller is a controller described by the following

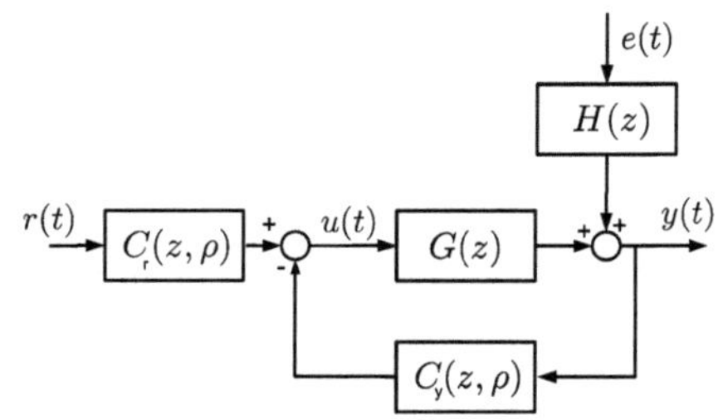

Fig. 7.1 Block-diagram of a 2DOF controller

control law

$$u(t) = C_r(z,\rho)r(t) - C_y(z,\rho)y(t) \tag{7.11}$$

where the "in-the-loop" controller $C_y(z,\rho)$ and the "pre-filter" controller $C_r(z,\rho)$ are causal transfer functions. This control structure is depicted in Fig. 7.1.

The same development presented in Sect. 7.1.1 leads to the following algorithm for the tuning of the controllers $C_y(z,\rho)$ and $C_r(z,\rho)$.

The IFT Algorithm for 2DOF Controllers At each step i of an iterative data-driven control design, estimate the gradient $\nabla J_{IFT}(\rho_i)$ through the following procedure.

1. collect data from the normal operation of the process; let the reference signal during this measurement period be denoted by $r_1(t)$ and the resulting controller's output and process output respectively by $u_1(t)$ and $y_1(t)$;
2. compute offline the signal $r_1(t) - y_1(t) \triangleq r_2(t)$;
3. perform an additional experiment applying $r_2(t)$ as the reference to the closed-loop system and measure the resulting controller's output and process output; let these measurements be denoted by $u_2(t)$ and $y_2(t)$ respectively;
4. collect once again data from the normal operation of the process, with the same reference signal as in the first step; let the resulting controller's output and process output be denoted respectively by $u_3(t)$ and $y_3(t)$;
5. compute

$$\widehat{\frac{\partial y(t,\rho_i)}{\partial \rho}} = \frac{1}{C_r(z,\rho_i)}\left[\left(\frac{\partial C_r(z,\rho_i)}{\partial \rho} - \frac{\partial C_y(z,\rho_i)}{\partial \rho}\right)y_3(t) + \frac{\partial C_y(z,\rho_i)}{\partial \rho}y_2(t)\right]$$

and

$$\widehat{\frac{\partial u(t,\rho_i)}{\partial \rho}} = \frac{1}{C_r(z,\rho_i)}\left[\left(\frac{\partial C_r(z,\rho_i)}{\partial \rho} - \frac{\partial C_y(z,\rho_i)}{\partial \rho}\right)u_3(t) + \frac{\partial C_y(z,\rho_i)}{\partial \rho}u_2(t)\right];$$

6. compute

$$\widehat{\nabla J(\rho_i)} = \frac{2}{N}\sum_{t=1}^{N}[y_d(t) - y_1(t)]\widehat{\frac{\partial y(t,\rho_i)}{\partial \rho}} + \lambda\frac{2}{N}\sum_{t=1}^{N}u_1(t)\widehat{\frac{\partial u(t,\rho_i)}{\partial \rho}}.$$

Note that this algorithm requires an extra measurement phase—Step 4 of the algorithm. But this is not a special experiment to be performed, only a measurement of data during normal operation of the process. With the algorithm above, including this additional measurement phase, the gradient estimate is again unbiased.

7.1.2.3 Optimizing Robustness

Consider the following H_2 objective function, which corresponds to $J_T(\rho)$ when $T_d(z) = 1$:

$$J_S(\rho) \triangleq J_T(\rho)\,|_{T_d(z)=1} = \bar{E}[r(t) - y(t,\rho)]^2.$$

Applying Parseval's Theorem, this objective function can also be written as

$$J_S(\rho) = \frac{1}{2\pi}\int_{-\pi}^{\pi} |S(e^{j\omega},\rho)|^2[\Phi_r(e^{j\omega}) + \Phi_v(e^{j\omega})]d\omega.$$

Observe that minimizing $J_S(\rho)$ is equivalent with minimizing the weighted H_2 norm of the sensitivity function $S(z,\rho)$, where the norm is weighted by the spectra of the reference and of the noise. But the norm of the sensitivity function $S(z,\rho)$ is directly related to the robustness of the closed-loop system, so that in a sense minimizing $J_S(\rho)$ (which is a particular case of $J_T(\rho)$) maximizes robustness. From a robustness point of view it would be more interesting to minimize the H_∞ norm, and we will take care of this aspect in a while. It is also useful in the context of robust control design to include an additional weighting function in the norm, in the form of a filter chosen by the designer. The cost function with this filter becomes

$$J_S(\rho) = \bar{E}[L_S(z)(r(t) - y(t,\rho))]^2.$$

It is usual in robust control theory to gather up the closed-loop sensitivities in a *sensitivity matrix* $\Sigma(z,\rho)$

$$\Sigma(z,\rho) = \begin{bmatrix} G(z)C(z,\rho)S(z,\rho) & G(z)S(z,\rho) \\ C(z,\rho)S(z,\rho) & S(z,\rho) \end{bmatrix} \tag{7.12}$$

and many robust control design methods consist in minimizing some weighted norm of this matrix. The sensitivity functions in this matrix can all be related to time-domain H_2 performance criteria in the same way as just described for $S(z,\rho)$. These relations, with frequency weighting filters included, are given below

$$\begin{aligned} J_{GCS}(\rho) &= \bar{E}[L_{GCS}(z)y(t,\rho)]^2 \\ &= \frac{1}{2\pi}\int_{-\pi}^{\pi} |L_{GCS}(e^{j\omega})|^2|S(e^{j\omega},\rho)|^2 \\ &\quad \times \{|G(e^{j\omega})C(e^{j\omega},\rho)|^2\Phi_r(e^{j\omega}) + \Phi_v(e^{j\omega})\}d\omega, \\ J_{CS}(\rho) &= \bar{E}[L_{CS}(z)u(t,\rho)]^2 \\ &= \frac{1}{2\pi}\int_{-\pi}^{\pi} |L_{CS}(e^{j\omega})|^2|C(e^{j\omega})S(e^{j\omega},\rho)|^2\{\Phi_r(e^{j\omega}) + \Phi_v(e^{j\omega})\}d\omega, \end{aligned}$$

$$
\begin{aligned}
J_{GS}(\rho) &= \bar{E}[L_{GS}(z)C^{-1}(z,\rho)y(t,\rho)]^2 \\
&= \frac{1}{2\pi}\int_{-\pi}^{\pi} |L_{GS}(e^{j\omega})|^2 |S(e^{j\omega},\rho)|^2 \\
&\quad \times \{|G(e^{j\omega})|^2 \Phi_r(e^{j\omega}) + |C^{-1}(e^{j\omega},\rho)|^2 \Phi_v(e^{j\omega})\} d\omega.
\end{aligned}
$$

The filters are important here, as they usually are in robustness designs, where their choice is strongly associated with the specified performance and robustness requirements. In this particular situation of IFT design for robustness, two aspects are important mentioning. First, the magnitude of the filters serves to weigh the different criteria against each other. Second, they allow to "transform" the H_2 norm into an approximation of the H_∞ norm, if we pick a band-pass filter around the maximum of $|S(z,\rho)|$. For this purpose, it was proposed in [14] a two step design. In the first step, $J_T(\rho)$ is optimized, and it is verified at which range of frequencies the sensitivities achieve their maximum values. Then this range of frequencies will be the passing band for the filters in the second step, where the robustness costs are included.

Unbiased estimates for the gradients of each one of these functions can be obtained using the unbiased estimate for the partial derivatives produced by IFT, in exactly the same way as the estimate of the gradient of $J_{IFT}(\rho)$. The gradients of these cost functions are given by the following expressions

$$
\begin{aligned}
\nabla J_S(\rho) &= 2\bar{E}\left[L_S C^{-1} u(t,\rho) L_S \left(\frac{\partial C^{-1}}{\partial \rho} u(t,\rho) + C^{-1}\frac{\partial u(t,\rho)}{\partial \rho}\right)\right] \\
\nabla J_{GCS}(\rho) &= 2\bar{E}\left[L_{GCS} y(t,\rho) L_{GCS} \frac{\partial y(t,\rho)}{\partial \rho}\right] \\
\nabla J_{CS}(\rho) &= 2\bar{E}\left[L_{CS} u(t,\rho) L_{CS} \frac{\partial u(t,\rho)}{\partial \rho}\right] \\
\nabla J_{GS}(\rho) &= 2\bar{E}\left[L_{GS} C^{-1} y(t,\rho) L_{GS} \left(\frac{\partial C^{-1}}{\partial \rho} y(t,\rho) + C^{-1}\frac{\partial y(t,\rho)}{\partial \rho}\right)\right].
\end{aligned}
$$

A complete objective function—say $J_R(\rho)$—would be one contemplating performance as specified by $J_T(\rho)$ and every aspect of robustness specified through the sum of each one of these objective functions

$$
J_R(\rho) = J_T(\rho) + J_{GCS}(\rho) + J_{CS}(\rho) + J_{GS}(\rho) + J_S(\rho).
$$

7.2 Frequency Domain Tuning

Leonardo Kammer, Bob Bitmead and Peter Bartlett presented in 2000 [8] a method for iterative tuning of a fixed structure controller aiming at optimizing the noise rejection performance. Their analysis is based on nonparametric models for the process and the controller. Given the nature of the computations involved, it becomes straightforward to include robustness measures.

7.2.1 Derivation

The performance criterion to be optimized is the noise rejection criterion $J_e(\rho)$, with an additional penalty in the control effort, that is, the following objective function

$$J_{FDT}(\rho) = J_e(\rho) + \lambda J_u(\rho) = \bar{E}[y^2(t,\rho) + \lambda u^2(t,\rho)]$$

with $\lambda \in \mathbb{R}^+$ a design parameter. It is assumed that there is no excitation at the reference: $r(t) \equiv 0$. Then the only external signal applied to the process is the noise and the signals appearing in the performance criterion are given by:

$$y(t,\rho) = S(z,\rho)v(t) \tag{7.13}$$

$$u(t,\rho) = -C(z,\rho)y(t,\rho) = -C(z,\rho)S(z,\rho)v(t). \tag{7.14}$$

Using Parseval's theorem the performance criterion can be written as follows

$$\begin{aligned} J_{FDT}(\rho) &= \frac{1}{2\pi}\int_{-\pi}^{\pi} [\Phi_y(e^{j\omega},\rho) + \lambda\Phi_u(e^{j\omega},\rho)]d\omega \\ &= \frac{1}{2\pi}\int_{-\pi}^{\pi} [1+\lambda|C(e^{j\omega},\rho)|^2]\Phi_y(e^{j\omega},\rho)d\omega. \end{aligned} \tag{7.15}$$

Calculating the gradient of $J_{FDT}(\rho)$ in (7.15) with respect to ρ yields

$$\begin{aligned} \nabla J_{FDT}(\rho) = \frac{1}{2\pi}\int_{-\pi}^{\pi} \Big\{ &[1+\lambda|C(e^{j\omega},\rho)|^2]\frac{\partial}{\partial\rho}\Phi_y(e^{j\omega},\rho) \\ &+ \Phi_y(e^{j\omega},\rho)\frac{\partial}{\partial\rho}[1+\lambda|C(e^{j\omega},\rho)|^2]\Big\} d\omega \end{aligned} \tag{7.16}$$

and the partial derivatives $\frac{\partial}{\partial\rho}[1+\lambda|C(e^{j\omega},\rho)|^2]$ and $\frac{\partial}{\partial\rho}\Phi_y(e^{j\omega},\rho)$ must be determined. The first one of these partial derivatives is given by

$$\frac{\partial}{\partial\rho}[1+\lambda|C(e^{j\omega},\rho)|^2] = 2\lambda\Re\left\{C^*(e^{j\omega},\rho)\frac{\partial C(e^{j\omega},\rho)}{\partial\rho}\right\}. \tag{7.17}$$

To determine the second one, determine the spectrum of $y(t,\rho)$ from (7.13):

$$\Phi_y(e^{j\omega},\rho) = \frac{|H(e^{j\omega})|^2}{|1+G(e^{j\omega})C(e^{j\omega},\rho)|^2}\sigma_e^2$$

whose gradient is

$$\frac{\partial}{\partial\rho}\Phi_y(e^{j\omega},\rho) = -2\Phi_y(e^{j\omega},\rho)\Re\left\{\frac{G(e^{j\omega})}{1+G(e^{j\omega})C(e^{j\omega},\rho)}\frac{\partial C(e^{j\omega},\rho)}{\partial\rho}\right\}. \tag{7.18}$$

Inserting (7.18) and (7.17) into (7.16) results in the final expression for the gradient:

$$\nabla J_{FDT}(\rho) = \frac{1}{\pi}\int_{-\pi}^{\pi} \Phi_y(e^{J\omega},\rho)\Bigg[\lambda\Re\left\{C^*(e^{J\omega},\rho)\frac{\partial C(e^{J\omega},\rho)}{\partial\rho}\right\} - (1+\lambda|C(e^{J\omega},\rho)|^2)\Re\left\{\frac{G(e^{J\omega})}{1+G(e^{J\omega})C(e^{J\omega},\rho)}\frac{\partial C(e^{J\omega},\rho)}{\partial\rho}\right\}\Bigg]d\omega. \tag{7.19}$$

In the gradient expression in (7.19) the controller's transfer function and its gradient ($C(e^{J\omega},\rho)$ and $\frac{\partial C(e^{J\omega},\rho)}{\partial\rho}$) are known a priori, and the spectrum $\Phi_y(e^{J\omega},\rho)$ is measured. The one term that is not available is the transfer function $\frac{G(e^{J\omega})}{1+G(e^{J\omega})C(e^{J\omega},\rho)}$, so in order to be able to calculate the gradient in (7.19) it is necessary to estimate this quantity. This can be done effectively by injecting a probe signal at the reference and using the following closed-loop relationship:

$$\frac{G(e^{J\omega})}{1+C(e^{J\omega},\rho)G(e^{J\omega})} = \frac{\Phi_{yr}(e^{J\omega})}{C(e^{J\omega},\rho)\Phi_r(e^{J\omega})} \tag{7.20}$$

where $\Phi_{yr}(e^{J\omega})$ is the cross-spectral density of the stochastic processes $y(t)$ and $r(t)$. The right hand side of this equation can be computed from the signals measured in this situation, which unfortunately is not the normal process' operation; an intrusive experiment is required.

Last, but not least, the Hessian of $J_{FDT}(\rho)$ can be computed similarly, and using the same measurements used to compute the gradient. Calculating the Hessian and appropriately grouping terms, it is found that the k,l element of the Hessian is given by (for readability we have omitted the dependence on ω and ρ in the presentation of this equation):

$$\begin{aligned}\frac{\partial^2 J_{FDT}^{k,l}(\rho)}{\partial\rho^2} = \frac{1}{\pi}\int_{-\pi}^{\pi}\Phi_y\Bigg[&\lambda\Re\left\{\frac{\partial C}{\partial\rho_l}\frac{\partial C^*}{\partial\rho_k} + C^*\frac{\partial^2 C^*}{\partial\rho_k\rho_l}\right\}\\ &-2\lambda\Re\left\{C^*\frac{\partial C}{\partial\rho_l}\right\}\Re\left\{\frac{G}{1+CG}\frac{\partial C}{\partial\rho_k}\right\}\\ &-2\lambda\Re\left\{C^*\frac{\partial C}{\partial\rho_k}\right\}\Re\left\{\frac{G}{1+CG}\frac{\partial C}{\partial\rho_l}\right\}\\ &-\Re\left\{\frac{G}{1+CG}\frac{\partial^2 C}{\partial\rho_l\rho_k} - 2\frac{G}{1+CG}\frac{\partial C}{\partial\rho_l}\frac{G}{1+CG}\frac{\partial C}{\partial\rho_k}\right.\\ &\left.-\left|\frac{G}{1+CG}\right|^2\frac{\partial C}{\partial\rho_l}\frac{\partial C^*}{\partial\rho_k}\right\}(1+\lambda|C|^2)\Bigg]d\omega.\end{aligned} \tag{7.21}$$

It is seen that indeed all the terms in this expression were already present in the gradient's formula, and hence no additional information is necessary. This allows the use of Newton-Raphson's optimization without extra experimental cost regarding the steepest descent.

The FDT gradient estimation procedure can be summarized as follows.

The FDT Algorithm At each step i of an iterative data-driven control design, estimate the gradient $\nabla J_{FDT}(\rho_i)$ through the following procedure.

1. collect data from the normal operation of the process with $r(t) \equiv 0$; let the output during this measurement period be denoted by $y_1(t)$;
2. estimate the spectrum of the output $y_1(t)$;

$$\hat{R}_y(\tau) = \frac{1}{N}\sum_{t=1}^{N} y_1(t)y_1(t-\tau)$$

$$\Phi_r(e^{j\omega}) = \frac{1}{N}\sum_{\tau=-\delta}^{\delta} w(\tau)\hat{R}_y(\tau)e^{-j\omega\tau}$$

   where $\omega(\cdot)$ is an appropriate window [13].
3. perform an additional experiment with a SRp reference $r(t)$ to the closed-loop system and measure the resulting output $y_2(t)$;
4. estimate the reference spectrum

$$\hat{R}_r(\tau) = \frac{1}{N}\sum_{t=1}^{N} r(t)r(t-\tau)$$

$$\Phi_r(e^{j\omega}) = \frac{1}{N}\sum_{\tau=-\delta}^{\delta} w(\tau)\hat{R}_r(\tau)e^{-j\omega\tau}$$

   and the cross-spectral density

$$\hat{R}_{yr}(\tau) = \frac{1}{N}\sum_{t=1}^{N} y_2(t)r(t-\tau)$$

$$\Phi_{yr}(e^{j\omega}) = \frac{1}{N}\sum_{\tau=-\delta}^{\delta} w(\tau)\hat{R}_{yr}(\tau)e^{-j\omega\tau}$$

   where $\omega(\cdot)$ is an appropriate window [13].
5. compute the sensitivity estimate from (7.20):

$$\frac{\widehat{G(e^{j\omega})}}{1+C(e^{j\omega},\rho_i)G(e^{j\omega})} = \frac{\Phi_{yr}(e^{j\omega})}{C(e^{j\omega},\rho_i)\Phi_r(e^{j\omega})}$$

6. compute $\nabla \widehat{J_{FDT}(\rho_i)}$ from (7.19) and $\nabla \widehat{J_{FDT}(\rho_i)}$ from (7.21) replacing $\frac{G(e^{j\omega})}{1+C(e^{j\omega},\rho_i)G(e^{j\omega})}$ by its estimate $\frac{\widehat{G(e^{j\omega})}}{1+C(e^{j\omega},\rho_i)G(e^{j\omega})}$ computed in Step 5.

7.2.2 Extensions

7.2.2.1 Guaranteeing Stability Along the Way

Given a controller for which the closed-loop is BIBO-stable, another controller will also yield a stable closed loop provided that it is "close enough" to the first controller. This reasoning results in a convenient test for stability to apply before changing the controller from one iteration to the next in a data-driven iterative control design, provided that the appropriate measure of "closeness" is used. The appropriate measure in this problem is the Vinnicombe distance between two controllers $\delta_\nu(C_1(z), C_2(z))$, which is defined as follows. For a given transfer function $Q(z)$, let $\eta(Q(z))$ be the number of poles outside the unit circle of this transfer function and $\zeta(Q(z))$ be the number of counterclockwise encirclements of the origin made by the Nyquist plot of $Q(z)$. If the following condition is satisfied

$$1 + C_1^*(z)C_2(z) \neq 0 \quad \& \quad \zeta(1 + C_1^*(z)C_2(z)) + \eta(C_2(z)) - \eta(C_1(z)) = 0$$

then

$$\delta_\nu(C_2(z), C_1(z)) = \left\| [1 + C_1(z)C_1^*(z)]^{-\frac{1}{2}}[C_1(z) - C_2(z)][1 + C_2(z)C_2^*(z)]^{-\frac{1}{2}} \right\|_\infty$$

else $\delta_\nu(C_2(z), C_1(z)) = 1$.

Think of the two controllers in this definition as the current controller at iteration i of the design, for which the system is currently operating in a stable manner, and of the second controller as the one that has just been determined and must be inserted in the loop. Before actually inserting it in the loop, we would like a guarantee that the resulting closed-loop will be stable. This will be guaranteed if the new controller represents a perturbation to the current closed-loop that is smaller than its stability margin.

Recall from the IFT presentation that robustness can be expressed by the norm of the matrix $\Sigma(z, \rho)$ in (7.12). An important measure of robustness of the closed-loop achieved with a given controller $C(z, \rho)$ is the following stability margin

$$b_{C(z,\rho)} \triangleq \|\Sigma(z, \rho)\|_\infty^{-1}. \tag{7.22}$$

With these measures of robustness of the closed-loop system at iteration i and of the distance between the current controller and the next one, we can use the following test to assert that the new controller will provide a BIBO-stable closed loop.

Theorem 7.1 *Let* $\rho_1 \in \Gamma$.[1] *If* $\delta_\nu(C(z, \rho_2), C(z, \rho_1)) < b_{C(z,\rho_1)}$ *then* $\rho_2 \in \Gamma$.

Testing this criterion requires calculating the Vinnicombe distance between the two controllers and the stability margin. These two can be calculated using the same

[1] That is, the closed-loop system with the controller $C(z, \rho_1)$ is BIBO-stable.

quantities already calculated in the FDT procedure. Indeed, the Vinnicombe distance only involves the transfer functions of the two controllers, and the stability margin can be calculated as

$$b_{C(z,\rho)} = \left\| \begin{matrix} \frac{\Phi_{ur}(e^{j\omega},\rho)}{C(e^{j\omega},\rho)\Phi_r(e^{j\omega},\rho)} & \frac{\Phi_{ur}(e^{j\omega},\rho)}{\Phi_r(e^{j\omega},\rho)} \\ \frac{\Phi_{yr}(e^{j\omega},\rho)}{C(e^{j\omega},\rho)\Phi_r(e^{j\omega},\rho)} & \frac{\Phi_{yr}(e^{j\omega},\rho)}{\Phi_r(e^{j\omega},\rho)} \end{matrix} \right\|_\infty^{-1}. \tag{7.23}$$

Clearly, Theorem 7.1 is valid for any two controllers, and certainly for any two subsequent controllers in any iterative tuning procedure. In principle, the same test can be applied to any iterative tuning procedure, not only in FDT and not only for this particular optimization criterion. But for other methods and/or performance criteria the quantities necessary for calculating the associated stability margins would have to be determined, maybe needing extra approximations and/or extra experiments. Whereas in FDT the stability test comes "for free", since all these quantities have already been calculated.

7.2.2.2 2DOF

Like IFT, FDT can also be extended to 2DOF controllers, that is, controllers described by (7.11). But unlike IFT, this does not require extra experiments or extra calculations. The procedure is still the same and the fact that the controller is 2DOF does not even change the equations used to calculate the gradient and the Hessian. Only the expressions (7.20) and (7.23), for estimating the sensitivity $\frac{G(e^{j\omega})}{1+G(e^{j\omega})C(e^{j\omega},\rho)}$ and for calculating the stability margin $b_{C(z,\rho)}$ respectively, are changed into the following

$$\frac{G(e^{j\omega})}{1+C_y(e^{j\omega},\rho)G(e^{j\omega})} = \frac{\Phi_{yr}(e^{j\omega})}{C_r(e^{j\omega},\rho)\Phi_r(e^{j\omega})}$$

$$b_{C(z,\rho)} = \left\| \begin{matrix} \frac{\Phi_{ur}(e^{j\omega},\rho)}{C_r(e^{j\omega},\rho)\Phi_r(e^{j\omega},\rho)} & \frac{\Phi_{ur}(e^{j\omega},\rho)C_y(e^{j\omega},\rho)}{\Phi_r(e^{j\omega},\rho)C_r(e^{j\omega},\rho)} \\ \frac{\Phi_{yr}(e^{j\omega},\rho)}{C_y(e^{j\omega},\rho)\Phi_r(e^{j\omega},\rho)} & \frac{\Phi_{yr}(e^{j\omega},\rho)C_y(e^{j\omega},\rho)}{\Phi_r(e^{j\omega},\rho)C_r(e^{j\omega},\rho)} \end{matrix} \right\|_\infty^{-1}.$$

7.3 Correlation-Based Tuning

Alireza Karimi, Ljubisa Mišković and Dominique Bonvin first presented the Correlation-based Tuning (CbT) method in their 2003 paper [9], and a careful account of its properties was later given in [10]. CbT aims at optimizing the reference tracking performance, but it does not do so by directly minimizing the objective function $J_y(\rho)$. Instead, the authors built their method upon an analogy with the correlation based methodologies in the system identification theory, and formulated the control design problem as one of identification of the ideal controller through instrumental variable methods. The CbT methodology is detailed in the sequel.

7.3.1 Derivation

Define, for convenience of notation, the following output error signal

$$\varepsilon(t,\rho) \triangleq y_d(t) - y(t,\rho) = [T_d(z) - T(z,\rho)]r(t) + S(z,\rho)v(t).$$

With the ideal controller $C_d(z)$ in the loop, $T(z,\rho) = T_d(z)$, and the output error would be $\varepsilon(t,\rho) = S(z,\rho)v(t)$; it would depend only on the noise, and not on the reference. So the ideal controller $C_d(z)$ is the one for which these two signals are not correlated, that is, for which the following equation is satisfied

$$\bar{E}[\varepsilon(t,\rho)r(t-\tau)] = \mathbf{0} \quad \forall\tau. \tag{7.24}$$

Therefore, the ideal controller's parameter ρ_d is the solution of the correlation equation (7.24) and it could in principle be found by solving this equation, but this is actually a system of infinite equations. In order to obtain a set of equations that can be solved, the system of equations (7.24) is truncated, keeping only those equations for small values of τ, which are the ones for which the correlation tends to be larger. Moreover, it can be advantageous to replace the reference $r(t)$ by another signal strongly correlated to it but still not correlated to the noise.

With this rationale in mind, define the instrumental variable $\xi(t) \in \mathbb{R}^q$ with $q \geq p$ such that $\bar{E}[v(t)\xi(t)] = \mathbf{0}$ and that all the elements of the vector $\bar{E}[r(t)\xi(t)]$ are nonzero. Then define the function $f(\cdot): \mathbb{R}^p \to \mathbb{R}^q$:

$$f(\rho) \triangleq \bar{E}[\varepsilon(t,\rho)\xi(t)].$$

The CbT method is based on finding the roots of this vector function, that is, in solving the following system of q nonlinear equations with p unknowns:

$$f(\rho) = \bar{E}[\varepsilon(t,\rho)\xi(t)] = \mathbf{0}. \tag{7.25}$$

If $q = p$, then the solution of this set of nonlinear equations can be found by an iterative algorithm, just the same as in the optimization problems treated all along this book, with an algorithm of the form:

$$\rho_{i+1} = \rho_i - \gamma_i R_i f(\rho_i). \tag{7.26}$$

Notice however the advantage of this algorithm over the previous methods. The right hand side of this equation does not depend on the gradient of some unknown function, but only on the function $f(\rho_i)$ itself. As a consequence, there is no need for intricate schemes for estimation of the gradient, or extra experiments; all that needs to be done is to estimate the function $f(\rho_i)$ by

$$\hat{f}(\rho_i) = \frac{1}{N}\sum_{t=1}^{N}\varepsilon(t,\rho_i)\xi(t). \tag{7.27}$$

There is, unfortunately, a caveat: the solution of the equation only guarantees that the tracking error is uncorrelated to the p samples of the reference that are closest to it in time, and not to the whole reference signal. Because p is usually very small, this motivates the use of larger numbers of equations ($q \gg p$), but then there are more equations than unknowns and the nonlinear system of equations will in general have no solution, unless Assumption B_y is satisfied. Nevertheless, this procedure has been successfully applied to practical examples [10], and it has shown some robustness against violations of Assumption B_y. A practical application will also be described in Chap. 8.

An alternative that is safer, but for which the above computational convenience of the CbT method is lost, is to "solve" the nonlinear system of equations in a least squares sense, that is, to minimize the following cost function

$$J_c(\rho) = f^T(\rho) f(\rho) = \bar{E}[\varepsilon(t,\rho)\xi(t)]^T \bar{E}[\varepsilon(t,\rho)\xi(t)]. \tag{7.28}$$

To optimize this cost function it is necessary to estimate its gradient, just as in the IFT and FDT methods. The gradient of the objective function $J_c(\rho)$ in (7.28) with respect to the parameter vector is given by

$$\nabla J_c(\rho) = 2\bar{E}\left[\frac{\partial \varepsilon(t,\rho)}{\partial \rho}\xi^T(t)\right]\bar{E}\left[\varepsilon(t,\rho)\xi(t)\right].$$

Similarly to the IFT case, an unbiased estimate of this gradient can be obtained by:

$$\widehat{\nabla J_c(\rho)} = 2\frac{1}{N}\sum_{t=1}^{N}\left[\frac{\partial \varepsilon(t,\rho)}{\partial \rho}\xi^T(t)\right]\frac{1}{N}\sum_{t=1}^{N}\left[\varepsilon(t,\rho)\xi(t)\right]. \tag{7.29}$$

Computing the estimate (7.29) requires measurements of the instrumental variable $\xi(t)$ (which is a nobrainer) and the computation of the partial derivative $\frac{\partial \varepsilon(t,\rho)}{\partial \rho}$. But $\varepsilon(t,\rho) = y(t,\rho) - y_d(t)$, that is:

$$\frac{\partial \varepsilon(t,\rho)}{\partial \rho} = \frac{\partial y(t,\rho)}{\partial \rho}.$$

Hence an estimate of the partial derivative $\frac{\partial \varepsilon(t,\rho)}{\partial \rho}$ can be obtained in exactly the same way as described in IFT, by means of the additional experiment described then. This provides an unbiased estimate of the gradient and thus convergence to a point where the gradient $\nabla J_c(\rho)$ is zero.

Another approach to obtain the derivative proposed by the CbT creators is to generate it based on a rough model for the process. If the control action is a linear regression

$$u(t,\rho) = \rho^T \phi(t)$$

with $\phi(t)$ a vector containing only past measurements of the control $u(t,\rho)$ itself and of the tracking error $r(t) - y(t,\rho)$, then the following expression is valid

$$\frac{\partial \varepsilon(t,\rho)}{\partial \rho} = \frac{B(z)}{P(z,\rho)}\phi(t,\rho) \tag{7.30}$$

where $B(z)$ and $P(z,\rho)$ are the numerator of the process transfer function and the denominator of the closed-loop transfer function $T(z,\rho)$ respectively. Then these two polynomial $B(z)$ and $P(z,\rho)$ can be approximated using a model for the process. This alternative way of estimating the gradient does not require the extra experiment but relies on a process model. The estimate will be unbiased only if the model estimate used for its calculation is unbiased, which is unlikely to happen. Hence, convergence can no longer be guaranteed with this alternative estimate of the gradient.

In short, the CbT algorithm can be described as follows. The description is given for the optimization version of the algorithm, in which $J_c(\rho)$ is minimized. The original and simpler version does not require any estimates and can be described simply by the recursion (7.26) and the calculation (7.27). On the other hand, the description below uses IFT-like estimate of the partial derivative of $y(t,\rho)$ with respect to ρ; for the alternative estimate of this quantity it suffices to eliminate step 3 and substitute accordingly step 4.

The CbT Algorithm—Optimization Version At each step i of an iterative data-driven control design, estimate the gradient $\nabla J_c(\rho_i)$ through the following procedure.

1. collect data from the normal operation of the process; let the reference signal during this measurement period be denoted by $r_1(t)$ and the resulting control input and process output respectively by $u_1(t)$ and $y_1(t)$;
2. compute offline the signal $r_1(t) - y_1(t) \triangleq r_2(t)$ and the instrumental variable vector $\xi(t)$;
3. perform an additional experiment applying $r_2(t)$ as the reference to the closed-loop system and measure the resulting control input and process output; let these measurements be denoted by $u_2(t)$ and $y_2(t)$ respectively;
4. compute

$$\widehat{\frac{\partial y(t,\rho_i)}{\partial \rho}} = \frac{1}{C(z,\rho_i)}\frac{\partial C(z,\rho_i)}{\partial \rho}y_2(t);$$

5. compute

$$\widehat{\nabla J_c(\rho_i)} = \frac{2}{N}\sum_{t=1}^{N}\widehat{\frac{\partial y(t,\rho_i)}{\partial \rho}}\xi^T(t)\sum_{t=1}^{N}\xi(t)[y_d(t) - y_1(t)].$$

As in most applications of IV ideas in different contexts, different choices of IVs have been proposed over the years of application experience with CbT. The simplest and more commonly found ones are the delayed samples of the reference itself and the noiseless simulated output with a rough model for the process (a choice of instrumental variable that has already been used in the VRFT method in Chap. 3).

7.4 Chapter Conclusions

Three different methods for obtaining the necessary quantities for optimization of the H_2 cost functions have been presented in this chapter. Each method can be applied to the optimization of a different control objective, that is, to a different cost function. In each method (except in the original version of CbT) it is necessary to collect data from a "special" experiment, that is, an experiment with a particular reference input that does not correspond to the normal operation of the process. These special (and inconvenient) experiments can be sometimes avoided by identifying a process model with data from the normal operation of the process, but then the convergence is no longer guaranteed because of the bias that will be present in this estimated model.

Notice that these procedures for estimation of the gradient and Hessian do not require that the controller is parameterized linearly as in Assumption LP. Assumption LP is used in most data-driven designs not for implementation of the algorithms, but rather for providing theoretical guarantees of convergence and to guide some design choices, as presented in the previous chapters of this book.

It is a very important practical issue in iterative data-driven control design to have some guarantee, at each iteration, that the next controller to be put in the loop is at least a stabilizing controller for the unknown process. The results previously presented in this book do not provide this guarantee—they tacitly assume that the (candidate) DOAs are contained in the stability set Γ. Even though this is a reasonable assumption, it should not be taken for granted, given the severe consequences that can result from putting a destabilizing controller in the loop. A test that provides in real-time a guarantee that the next controller is a stabilizing one is provided in FDT, as presented in Sect. 7.2. This stability test has been developed into a simpler and more effective form in [7]. Given its relevance, this subject has been given a broader approach later on, and a theoretical framework of its own has been developed, resulting in tests that can be applied equally well with any data-driven control design methodology [2].

References

1. F. de Bruyne, L.C. Kammer, Iterative feedback tuning with guaranteed stability, in *American Control Conference*, vol. 21, San Diego, CA, USA, 1999, pp. 3317–3321
2. A. Dehghani, A. Lecchini, A. Lanzon, B. Anderson, Validating controllers for internal stability utilizing closed-loop data. IEEE Trans. Autom. Control **59**(11), 2719–2725 (2009)
3. R. Hildebrand, A. Lecchini, G. Solari, M. Gevers, Asymptotic accuracy of iterative feedback tuning. IEEE Trans. Autom. Control **50**(8), 1182–1185 (2005)
4. H. Hjalmarsson, Iterative feedback tuning – An overview. Int. J. Adapt. Control Signal Process. **16**(5), 373–395 (2002)
5. H. Hjalmarsson, M. Gevers, S. Gunnarsson, O. Lequin, Iterative feedback tuning: Theory and applications. IEEE Control Syst. Mag. **18**(4), 26–41 (1998)
6. H. Hjalmarsson, S. Gunnarsson, M. Gevers, A convergent iterative restricted complexity control design scheme, in *33rd IEEE Conference on Decision and Control*, Lake Buena Vista, USA, 1994

7. L.C. Kammer, Stability assessment for cautious iterative controller tuning. Automatica **41**, 1829–1834 (2005)
8. L.C. Kammer, R.R. Bitmead, P.L. Bartlett, Direct iterative tuning via spectral analysis. Automatica **36**, 1301–1307 (2000)
9. A. Karimi, L. Mišković, D. Bonvin, Iterative correlation-based controller tuning: Application to a magnetic suspension system. Control Eng. Pract. **11**, 1069–1078 (2003)
10. A. Karimi, L. Mišković, D. Bonvin, Iterative correlation-based controller tuning. Int. J. Adapt. Control Signal Process. **18**, 645–664 (2004)
11. A. Lecchini, M. Gevers, On iterative feedback tuning for non-minimum phase plants, in *41st IEEE Conference on Decision and Control*, 2002, pp. 4658–4663
12. A. Lecchini, M. Gevers, J. Maciejowski, An iterative feedback tuning procedure for loop transfer recovery, in *IFAC Symposium on System Identification*, Newcastle, Australia, 2006
13. L. Ljung, *System Identification – Theory for the User*, 2nd edn. (Prentice Hall, New York, 1999)
14. H. Procházka, M. Gevers, B.D.O. Anderson, C. Ferrera, Iterative feedback tuning for robust controller design and optimization, in *IEEE Conference on Decision and Control – European Control Conference*, Seville, Spain, 2005
15. G. Solari, M. Gevers, Unbiased estimation of the Hessian for iterative feedback tuning (IFT), in *43th Conference on Decision and Control*, Atlantis, Paradise Island, Bahamas, 2004, pp. 1759–1760

Chapter 8
Experimental Results

We have given several simulation examples along the various chapters of this book to better explain the concepts and the methods, but it is hard to illustrate the actual application of a model-free design with simulation examples. In simulation, we always know everything there is to know about the process, and it may sound artificial and arbitrary to determine which part of this information will be used for the design and which will not. Only with practical applications can we hope to clarify how a data-driven design is really performed. Accordingly, in this chapter we present the practical application of data-driven control design to real systems.

The designer must make the problem assessment and the corresponding design decisions using only whatever rudimentary and/or imprecise information about the process is available *a priori* to her/him. These assessments and decisions may involve the choice of a reference model, an estimate of to which extent Assumption B_y is violated, the necessity for and proper setting of cost function shaping, the choice of the search direction in the iterative optimization, and/or several other aspects discussed along this book. The data collected from these systems are affected by noise, whose level and spectrum are not known *a priori* and can vary widely, as well as by nonlinearities. With the presentation of these practical applications we expect to make clearer to the reader how the theory translates into the real world, what kind of information is typically available *a priori* and how the designer should make his/her choices based on this information. Besides, of course, showing evidence that the theory actually works and that the design for any specific process can be automated.

We present the control of three different processes: the flow control of water in a system of tanks, the speed control of a DC motor, and the control of the temperature of a heated resistance. Each process is representative of typical processes found in industry and possesses distinct features that will illustrate distinct aspects of data-driven control applications.

In general, data-driven methods aim to minimize the cost function

$$J_T(\rho) = J_y(\rho) + J_e(\rho),$$

A. Sanfelice Bazanella et al., *Data-Driven Controller Design*,
Communications and Control Engineering,
DOI 10.1007/978-94-007-2300-9_8,

where the reference tracking term $J_y(\rho)$ is due to the reference signal only and the noise rejection term $J_e(\rho)$ is due only to the noise. As seen along the book, some data-driven design methods minimize only the term $J_y(\rho)$, other methods minimize only $J_e(\rho)$, and some methods use the composite cost function $J_T(\rho)$. We will present designs with three different data-driven methods: Virtual Reference Feedback Tuning (VRFT), Iterative Feedback Tuning (IFT) and Correlation based Tuning (CbT). Recall that the VRFT and CbT methods minimize the reference tracking term $J_y(\rho)$ while the IFT method minimizes the composite cost function $J_T(\rho)$. As each method minimizes a different cost function, one cannot expect that they will result in the same controller parameters or control performance. For this reason alone, the controllers obtained with each one of them will be different, and comparisons among different methods must be made and interpreted very carefully.

Because only a finite set of data will be available and the cost functions involve expectations, they can not be calculated exactly, but only estimated from this finite set of data. The function that can actually be computed in a general case is the following estimate of the composite cost $J_T(\rho)$:

$$\hat{J}_T(\rho) = \frac{1}{N}\sum_{t=1}^{N}(y(t,\rho) - y_d(t))^2. \tag{8.1}$$

Note that, by definition:

$$J_T(\rho) = \lim_{N\to\infty} E[\hat{J}_T(\rho)].$$

Moreover, to express the quality of any given design we prefer to work with the square root of this function, because then the function has the same physical unit as the output and thus its value has a more evident physical interpretation. The performance of a control loop will thus be expressed by the following criterion:

$$\check{J}_T(\rho) = \sqrt{\hat{J}_T(\rho)}. \tag{8.2}$$

Approximations can—and will—be obtained for the reference tracking cost $J_y(\rho)$ by computing (8.1) with a filtered version of the output. The details of these computations are given in the next section.

8.1 A Liquid Flow Process

The first practical example to be presented is the control of water flow in a system of two tanks. The schematic diagram in Fig. 8.1 describes the main parts of the process. The whole process is built with of-the-shelf industrial equipments (pump, valve, sensors and tanks).

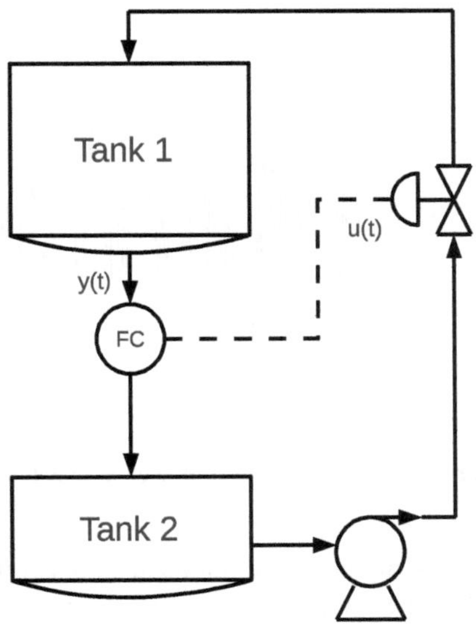

Fig. 8.1 Schematic diagram of the water flow control system. The water is pumped up from tank 2 to tank 1 through a valve and flows back to tank 2 by gravity. The output flow of tank 1 is the process variable $y(t)$ and the valve opening is the manipulated variable $u(t)$

The objective of the control system is to control the output flow of one of the tanks (the output variable $y(t)$), which is accomplished through the manipulation of the opening of a valve which is at the input of this same tank (the input variable $u(t)$). The flow entering the tank changes its level, which changes the pressure at the bottom of the tank and consequently the output flow. The output signal is the flow measure, given in milliliters per second ($\frac{\text{ml}}{\text{s}}$). As usual in flow measurement, the output signal is quite noisy. This measurement noise occurs in a wide range of frequencies, including the same range where the signals are observed, which makes it difficult to perform an effective filtering. The control system has some constraints, the most relevant of which being that only PI controllers can be implemented.

We begin with no knowledge whatsoever about the process' behavior and proceed with obtaining a minimal amount of information that will allow us to set up the data-driven design. Let us start by assessing the noise level of the process. This will serve as a benchmark for the achievable values of the cost functions, because if the noise where white then this would be the minimum value of $J_e(\rho)$.[1] Notice that this is probably not the minimum value of $J_e(\rho)$, because it is unlikely that the noise will actually be white. Still, it gives a ballpark indication of the achievable values for the performance indexes. If a control design results in a closed-loop system such that the cost evaluates at several times this value, it will most likely be a poor design that can be much improved; on the other hand, it is pointless to try and make the cost significantly smaller than this value.

Proceeding with the noise level estimate, an open-loop experiment is performed with a constant opening of the valve, at 85% of its full span. We collected $N =$

[1] Recall the discussion on the minimum variance controller in Chap. 2.

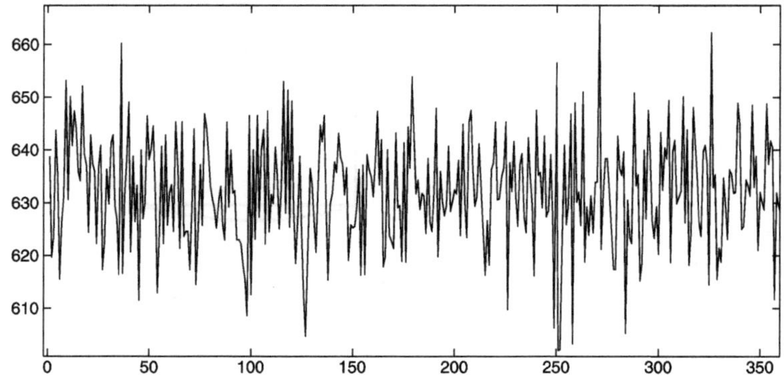

Fig. 8.2 Flow of an open-loop experiment, used to estimate $J_e(\mathbf{0})$

360 samples of the output with a sampling interval of 10 seconds and the result of this one hour experiment is presented in Fig. 8.2. The data collected were used to estimate the noise level of this process, and this calculation yields

$$\check{J}_e(\mathbf{0}) = 10.5215 \text{ ml/s} = \sqrt{\bar{E}[\nu(t)]^2}.$$

We would like the closed-loop system to behave as specified by the first order reference model

$$T_d(z) = \frac{1-a}{z-a},$$

where $a \in (0, 1)$ is the pole of the closed-loop system, to be specified. As said before, we can only implement PI controllers in the system, which can be described by

$$C(z, \rho) = \frac{\varrho_1 z + \varrho_2}{z-1}.$$

Assumption B_y is satisfied if and only if

$$G(z) = \frac{b}{z-c}$$

for some real numbers b and c. Assumption B_y is satisfied to exactly the same extent, and under exactly the same conditions, that such a first-order model appropriately describes the process dynamics; both statements are equivalent. This first-order model may or may not be a reasonable model from first principles analysis of the process, depending on the particular dimensions of the process, its operating point and the particular adjustment of the process' physical characteristics (the speed of the pump, the setting of the valve positioning device, etc.). Without any further modeling and analysis, one must assume that the process model is more complex than a first-order transfer function and, as a consequence, that Assumption B_y is not satisfied.

In the sequel the parameters of the PI controllers will be tuned by direct data-driven design (VRFT) and an iterative data-driven design (IFT).

8.1.1 Direct Method

In order to obtain data to apply the instrumental variables version of the VRFT method, two open-loop experiments were performed. In these experiments, the input signal is a step change from 80% to 85% in the opening of the valve. The output of one of these experiments is presented in Fig. 8.3 and the input applied to the process is shown in Fig. 8.4. Notice, from Fig. 8.3, that the signal-to-noise ratio is low; this will probably impose some difficulties to the data-driven method designs, particularly for those who minimize only the $J_y(\rho)$ term. In spite of the high level of noise, the VRFT with LS estimation will also be used for illustration purposes.

It is possible to estimate the settling time of the process at around 70 samples, which gives us the information that a first-order approximation of the process' transfer function would have a dominant open-loop pole close to 0.94. This is an useful information to decide the value of the pole of the closed-loop reference model $T_d(z)$. We chose to design the controller for three different reference models: with a equal to 0.80, 0.85 and 0.90. All of these reference models represent responses that are

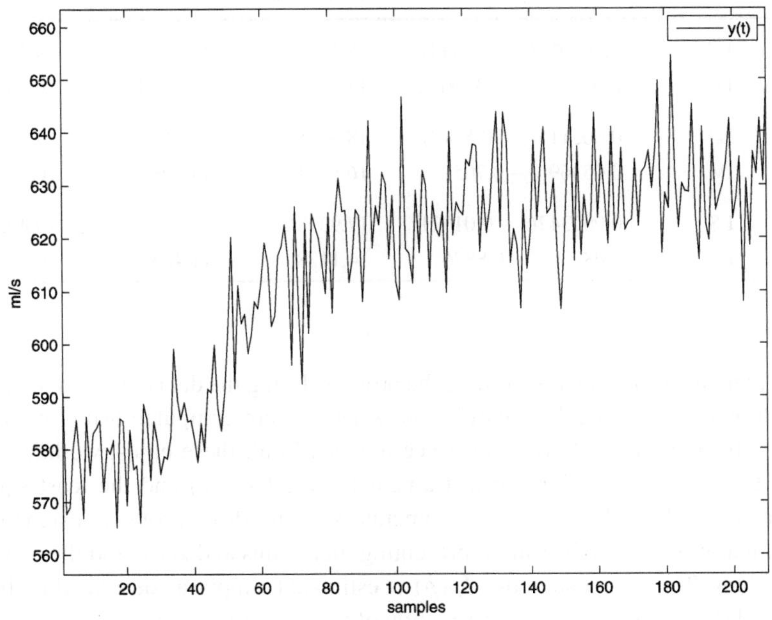

Fig. 8.3 Output signal of the open-loop experiment in the liquid process

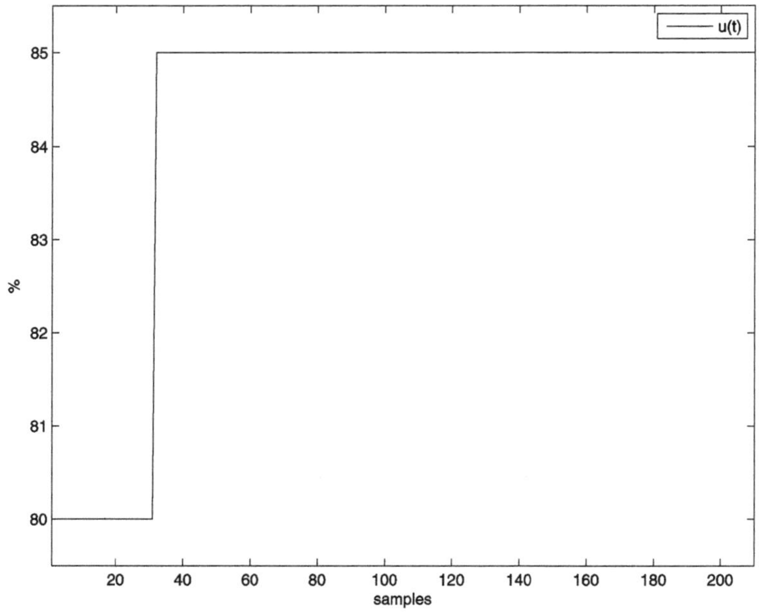

Fig. 8.4 Input signal of the open-loop experiment in the liquid process

Table 8.1 Flow controllers obtained with VRFT for different reference models

a	Method	$\text{num}(C(z,\hat{\rho}))$	$\check{J}_T(\hat{\rho})$ [ml/s]	$\check{J}_y(\hat{\rho})$ [ml/s]	$\check{J}_e(\hat{\rho})$ [ml/s]
0.80	LS	$0.003970(z+0.7444)$	18.6139	18.0556	9.2721
	IV	$0.03166(z-0.3300)$	17.1334	15.7430	10.6531
0.85	LS	$0.005115(z+0.5117)$	18.8208	17.8601	9.9749
	IV	$0.05369(z-0.7057)$	16.8571	14.7268	10.9266
0.90	LS	$0.007342(z-0.01021)$	17.9318	16.5554	10.4929
	IV	$0.08920(z-0.8876)$	14.1128	11.4537	10.6317

faster than the open-loop response of the process. Using the data collected in the two open-loop experiments, the controller parameters were computed using the VRFT method for the three different reference models. In all these cases, we have computed the parameters using the instrumental variables (IV) and the least squares (LS) method. Table 8.1 shows the numerators of the PI controllers designed (the denominator is always the same), presenting their gains and zeros and the resulting cost values. The composite cost $\check{J}_T(\rho)$ is estimated simply as described earlier in (8.1) and (8.2), but the separate estimation of the noise rejection cost ($\check{J}_e(\rho)$) and of the reference tracking cost ($\check{J}_y(\rho)$) requires some extra attention.

Recall that the reference tracking and noise rejection performance criteria are defined respectively as

$$J_y(\rho) = \bar{E}[y_r(t,\rho) - y_d(t)]^2,$$

$$J_e(\rho) = \bar{E}[y_e(t,\rho)]^2,$$

where $y_r(t,\rho) = T(z,\rho)r(t)$ and $y_e(t,\rho) = S(z,\rho)v(t) = y(t,\rho) - y_r(t,\rho)$. Neither $y_r(t,\rho)$ nor $y_e(t,\rho)$ are measurable, so $y_r(t,\rho)$ is approximated by a low-pass filtered version—$y_f(t,\rho)$—of the measured output $y(t,\rho)$. Then the two costs are estimated by

$$\check{J}_y(\hat{\rho}) = \sqrt{\frac{1}{N}\sum_{t=1}^{N}(y_f(t,\hat{\rho}) - y_d(t))^2}$$

$$\check{J}_e(\hat{\rho}) = \sqrt{\frac{1}{N}\sum_{t=1}^{N}(y(t,\hat{\rho}) - y_f(t,\hat{\rho}))^2}.$$

Now, from the very definition of the three costs, it is expected that the following relationship should be satisfied.

$$\check{J}_y^2(\hat{\rho}) + \check{J}_e^2(\hat{\rho}) = \check{J}_T^2(\hat{\rho}).$$

However, the estimations of the three performance criteria are imprecise due to the finiteness of the data. Moreover, the estimation of the noise rejection and of the reference tracking performance criteria are also affected by the approximation $y_r(t,\rho) \approx y_f(t,\rho)$ that has been necessary for their calculation. As a result, this identity is not exactly verified in the data presented in Table 8.1: a difference around 10% is observed in all the lines of this table. This small but not negligible discrepancy must be bore in mind when analyzing the results presented in Table 8.1.

For all different values of a, the use of instrumental variables (IV) has resulted in a better performance when compared to the Least Squares (LS) solution. A fair comparison between two VRFT designs should take into account only the reference tracking criterion, since this is the performance criterion that VRFT optimizes. This comparison shows that the LS design has resulted in values of the performance criterion $\check{J}_y(\hat{\rho})$ that are significantly higher (that is, worse) than the IV designs for the three different reference models with $a = 0.9$ (44.5% higher), $a = 0.85$ (21.2%) and $a = 0.8$ (12.8%). On the other hand, the lowest cost values were obtained using $a = 0.90$. This is expected because this is the slowest reference model, and thus the easiest to be followed—it is the one for which the violation of Assumption B_y is the least substantial.

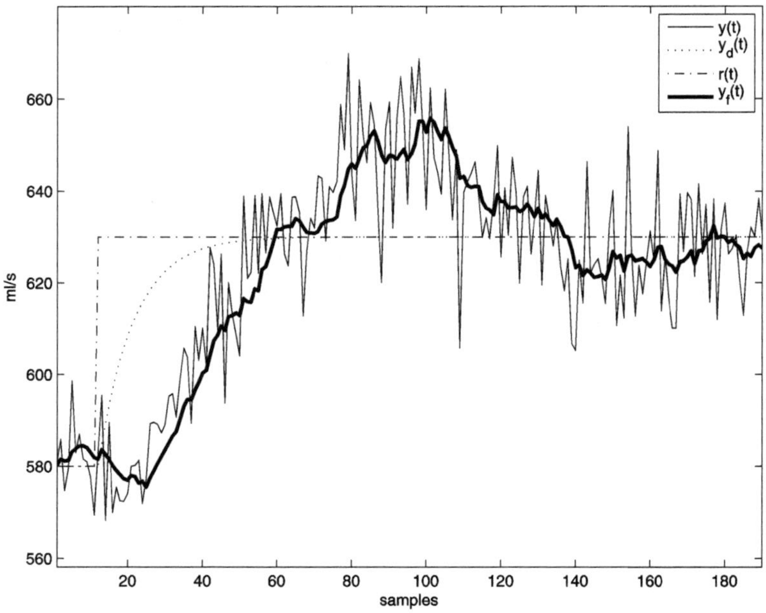

Fig. 8.5 Closed-loop response of the system with the PI controller obtained with VRFT using LS for a reference model response with $a = 0.90$

A visualization of the results obtained for the reference model with $a = 0.90$ is provided in Figs. 8.5–8.8, illustrating the results shown in Table 8.1. The controller computed with the VRFT using the Least Squares estimate for this case is

$$C_{LS}(z,\hat{\rho}) = \frac{0.007342(z - 0.01021)}{z - 1}.$$

Figures 8.5 and 8.6 present the output and the control signal, respectively, of the closed-loop system with this controller. The filtered output $y_f(t)$ used to compute the individual cost functions as described before is also shown. We can see that the closed-loop behavior is quite different from the desired reference model. The estimated cost of the controller obtained with the LS was $\check{J}_T(\rho) = 17.9318$ ml/s.

The controller computed with the VRFT method using the instrumental variables estimate is

$$C_{IV}(z,\hat{\rho}) = \frac{0.08920(z - 0.8876)}{z - 1}.$$

Figures 8.7 and 8.8 present the output and the control signal, respectively, of the resulting closed-loop system. The response obtained with IV is better than the one obtained using LS, but still far from the desired response $y_d(t)$, resulting in a cost $\check{J}_T(\hat{\rho}) = 14.1128$ ml/s. Taking into account that the cost of the noise in the open-loop experiment was $\check{J}_e(\mathbf{0}) = 10.5215$ ml/s, a similar value for $\check{J}_T(\rho)$ could be expected at the optimum controller, along with $\check{J}_y(\rho) \approx 0$. Because Assumption B_y is

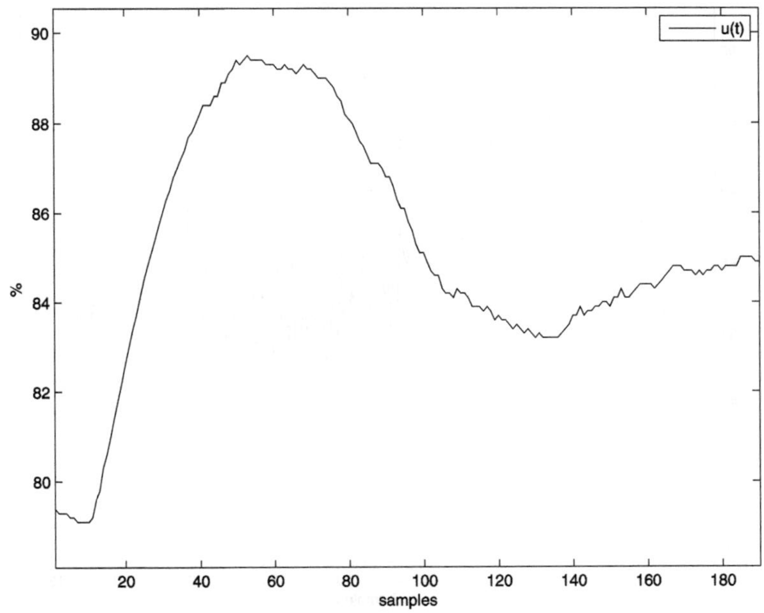

Fig. 8.6 Control signal of the PI controller obtained with VRFT for a reference model with $a = 0.90$, using LS

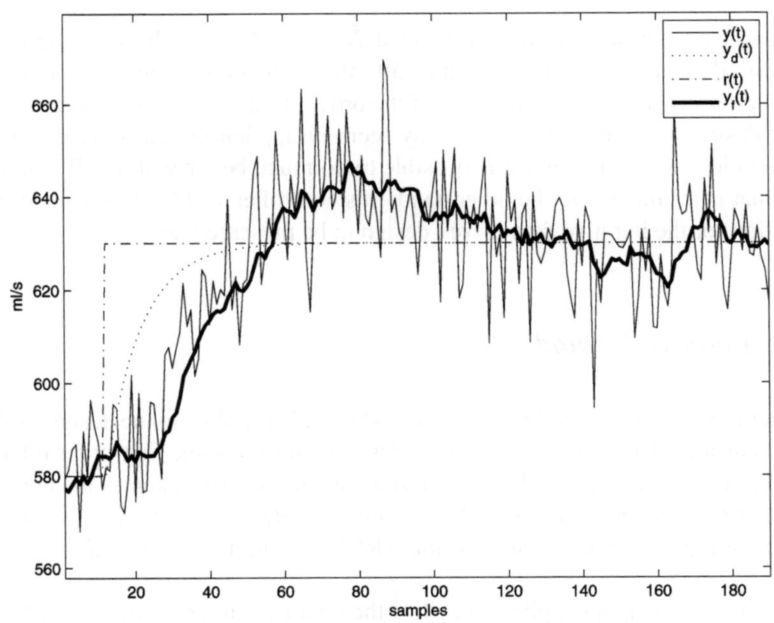

Fig. 8.7 Closed-loop response of the system with the PI controller obtained with VRFT using IV for a reference model response with $a = 0.90$

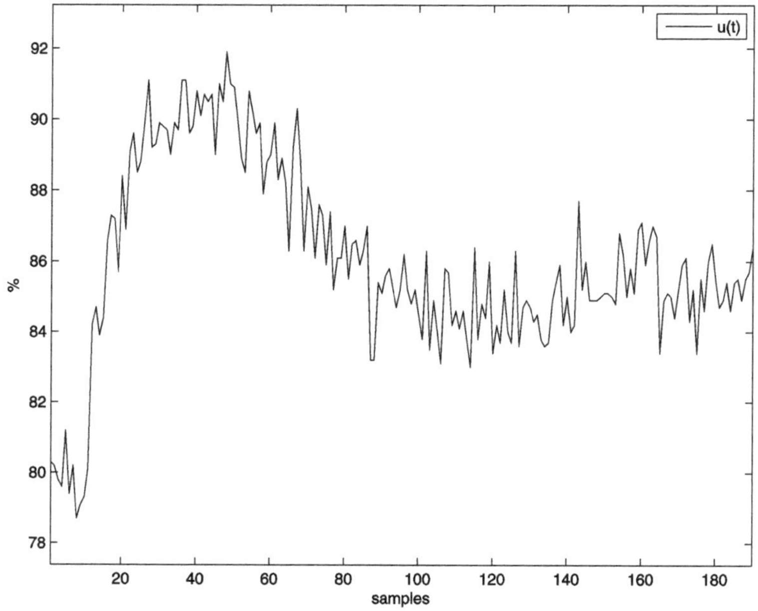

Fig. 8.8 Control signal of the PI controller obtained with VRFT for a reference model with $a = 0.90$, using IV

not satisfied, the best controller has yielded $\check{J}_y(\rho) = 11.4537$ ml/s. Considering that the desired behavior was a variation of 50 ml/s in the output, one can say that the output presents an average difference of the order of 22.9% ($= \frac{11.4537}{50}$) with respect to the desired behavior. This result may seem disappointing, but it should not be: it is not clear whether or not it is possible to do much better with the PI controller class that is available. On the other hand, this controller will be a very good initial condition for the iterative data-driven design to be presented next.

8.1.2 Iterative Method

The Iterative Feedback Tuning (IFT) method was also applied to the design of the PI flow controller. The controller has been designed for the same first-order reference model as before, with $a = 0.9$, and the initial controller was taken as the controller designed for this case with the VRFT-IV method. Recall that the IFT design does not aim at the same performance as the VRFT design; it minimizes $J_T(\rho)$ instead of $J_y(\rho)$.

The reference signal applied to collect the data used to compute the gradient of the cost function is the following

$$r(t) = 605 + 25sq\left(\frac{2\pi t}{360}\right),$$

Table 8.2 Controller evolution with IFT

i	$\text{num}(C_I(z, \rho_i))$	$\check{J}_T(\rho_i)$ [ml/s]	$\check{J}_y(\rho_i)$ [ml/s]	$\check{J}_e(\rho_i)$ [ml/s]
0	$0.0892z - 0.07918$	13.5050	11.2810	10.4293
1	$0.0919z - 0.08039$	12.5571	9.9179	10.4086
2	$0.0952z - 0.08134$	12.0205	8.8239	10.4167
3	$0.0970z - 0.08203$	12.5843	9.8802	10.4358
4	$0.0988z - 0.08319$	13.3915	9.6905	11.6510
5	$0.1048z - 0.08483$	9.8159	8.4707	7.7586

and the initial controller used in the iterative procedure is the one obtained with the VRFT method using instrumental variables. The steepest descent optimization was used, and in order to compute the step size sequence using the second method presented in Sect. 5.4.3, the following approximate model of the process was used

$$G(z) = \frac{0.6465}{z - 0.9573}.$$

This model was estimated from the same data collected for the application of the VRFT method.

The evolution of the controller parameters for five iterations of the IFT design is presented in Table 8.2. It is important to realize that the cost values presented in this Table are calculated for the square-wave reference being applied, and not for a reference step as the results presented previously in the direct design. This is why the costs in the first line of Table 8.2 are different from the ones presented in Table 8.1 for the same controller.

After five iterations the method achieved the controller

$$C(z, \rho_5) = \frac{0.1048z - 0.08483}{z - 1}$$

and the cost was reduced to $J_T(\rho_5) = 9.8159$ ml/s. Figure 8.9 shows the process variable in response to a reference step that is obtained in closed-loop with this controller, while Fig. 8.10 shows the corresponding controller's output. Taking into account that the noise level estimated in open loop was $\check{J}_T(\mathbf{0}) = \check{J}_e(\mathbf{0}) = 10.5215$ ml/s, the resulting performance with the final controller is quite satisfactory, as it yields a smaller cost value.[2]

[2]Notice, however, that there is a random component in the cost value, since each time it is measured a different noise realization is present. So, the low cost value obtained in the last iteration may not be as low if measured again.

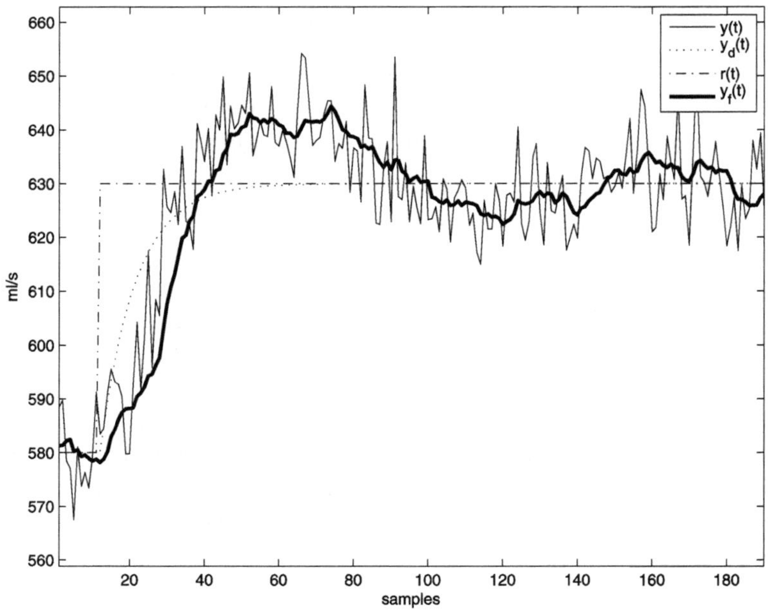

Fig. 8.9 Closed-loop response of the system with the PI controller obtained with IFT after five iterations for a reference model response with $a = 0.90$

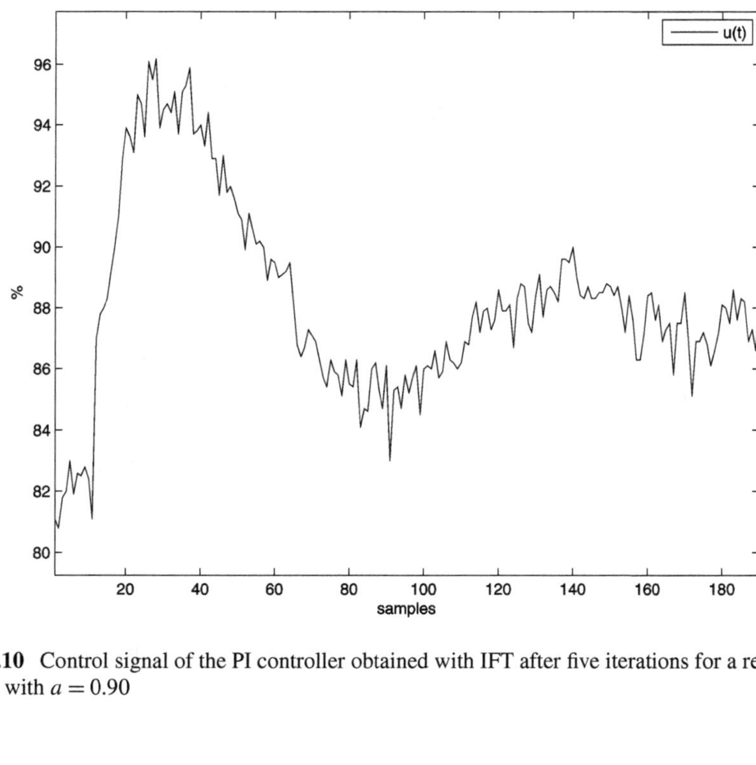

Fig. 8.10 Control signal of the PI controller obtained with IFT after five iterations for a reference model with $a = 0.90$

8.2 A DC Motor

The second experiment to be presented is the speed control of a DC motor, using the educational kit described in [1]. The output of the process $y(t)$ is the motor's speed, which is measured by an encoder and given in radians per second, whereas the manipulated variable $u(t)$ is the voltage applied to the motor, given in volts. The sampling period chosen for this application is 0.01 seconds, and in the results to be presented the time variable is always expressed in number of samples instead of seconds.

A PI controller is available, with the following structure

$$C(z,\rho) = [k_p \;\; k_i] \begin{bmatrix} 1 \\ \frac{z}{z-1} \end{bmatrix}, \tag{8.3}$$

where k_p is the proportional gain and k_i is the integral gain.

A DC motor is a fairly standard process, both in textbooks and in industrial practice. For speed control purposes it is usually modeled by a second order transfer function, which represents its two main dynamics: the mechanical part and the electrical part. The electrical dynamics tend to be much faster than the mechanical dynamics, which often motivates the use of a simplified first order model, but we will not make this simplification here. Let us work under the assumption that the speed control process is appropriately represented by a second-order model, that is, that the process' transfer function is of the form

$$G(z) = k_g \frac{z}{(z-b)(z-c)} \tag{8.4}$$

for some positive real numbers k_g, $b < 1$ and $c < 1$. With the process (8.4) and the controller (8.3), the closed-loop transfer function will be

$$T(z,\rho) = \frac{Kz(z-f)}{(z-b)(z-c)(z-1) + Kz(z-f)} \tag{8.5}$$

where we have defined $K = k_g(k_i + k_p)$ and $f = \frac{k_p}{k_i + k_p}$ for convenience of notation. The order of the closed-loop transfer function is, in general, three, but it is reduced to two when either $f = c$ or $f = b$; in any case, it is always larger than or equal to two.

The reference model will be chosen taking into account this basic knowledge about the process, trying to enforce Assumption B_y. We will analyze two situations:

1. the reference model has order one, which implies that the desired behavior cannot be achieved with the PI controller; this is the *mismatched case*;
2. the reference model has order two and the desired behavior can be achieved with the PI controller; this is the *matched case*.

In the sequel we will design controllers for these two situations using both a direct design method and an iterative design method.

8.2.1 Direct Method

Let us apply the VRFT method to design the PI controller. Even though the noise level is low in this particular application, still the instrumental variable method was applied to minimize $J^{VR}(\rho)$; the repeated experiment IV was used, with two open-loop experiments. The same input was utilized in both experiments: a square signal described by

$$u(t) = 3 + sq\left(\frac{2\pi t}{333}\right). \tag{8.6}$$

Let us open a brief parenthesis on the practical meaning of signal richness. A square wave is, in theory, sufficiently rich of infinite order. But the magnitude of the harmonics decay linearly with their order: the third harmonic is three times smaller than the first harmonic, the fifth is five times smaller, and so on. As the orders of the harmonics increase, their magnitudes become negligible with respect to the first harmonic, so for any practical purposes it would not be appropriate to count these negligible harmonics in the richness of the signal. How small should be considered negligible in a given practical case depends on its specificities—the noise level, the precision required for the solution, etc. An attitude in the very conservative side is to neglect all the harmonics but the first one, and work as if the square-wave were SR2; then the signal above is considered to be SR3, which is more than enough in this particular case because the controller's parameter is of dimension two. Let us now close this parenthesis.

Two experiments with the same input (8.6) have been performed; the results are presented in Figs. 8.11 and 8.12 (the experiments are visually indistinguishable from each other because the noise level is very low, so only the result of one experiment

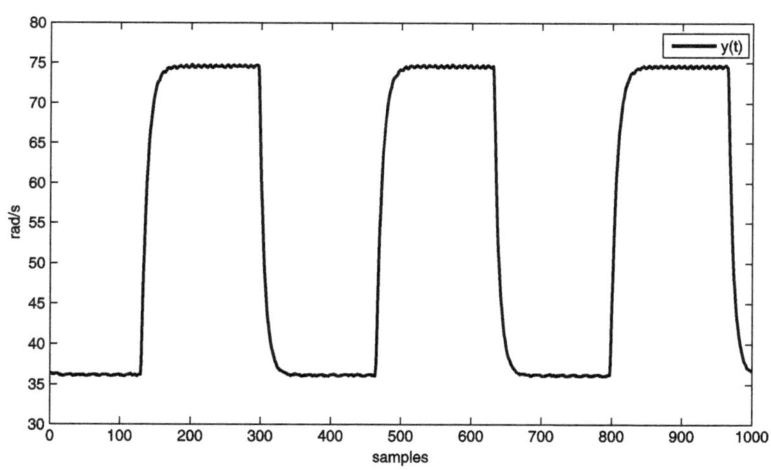

Fig. 8.11 Output collected from the open-loop experiment in the DC motor

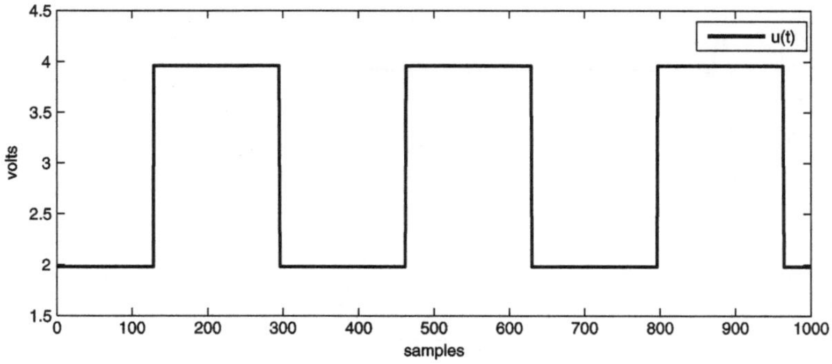

Fig. 8.12 Input collected from the open-loop experiment in the DC motor

has been plotted). We have collected $N = 1{,}000$ samples in order to perform the controller parameters' estimate.

8.2.1.1 Mismatched Case

Let us choose (once again) the simplest reference model, described by the first order transfer function

$$T_d(z) = \frac{1-a}{z-a}, \tag{8.7}$$

where a is the closed-loop pole and the transfer function gain is such that the steady-state gain is one, that is, $T_d(1) = 1$. Although the structure of the controller is given by (8.3), the controllers will be presented in the zero-pole form, in order to facilitate the visualization and the analysis of the results.

From the open-loop experiment performed, which has been presented in Fig. 8.11, the settling time of the process can be estimated at approximately 40 samples. Incidentally, this corresponds to a process with a transfer function possessing a dominant pole at $b \approx 0.9$ (dominant meaning that $|c|$ is significantly smaller than $|b|$).

Let us specify a performance with settling time of approximately 10 samples, which corresponds to a reference model with $a = 0.7$. This is a bold choice, as the closed-loop is asked to be about four times faster than the process in open-loop. Assumption B_y is not satisfied for any value of a, but the reference model becomes farther away from the achievable performance as a becomes closer to zero; otherwise phrased, the violation of Assumption B_y is more severe for faster reference models.[3] So, this value of a may not be a smart choice of reference model, but let us see what happens anyways.

[3]A root locus argument would show that, and this argument requires no knowledge of the process other than the assumption that its transfer function is of the form (8.4).

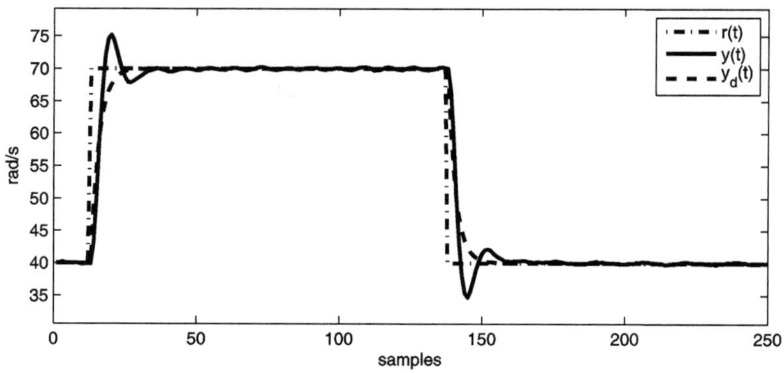

Fig. 8.13 Closed-loop responses of the reference model with $a = 0.7$ and the correspondent PI controller

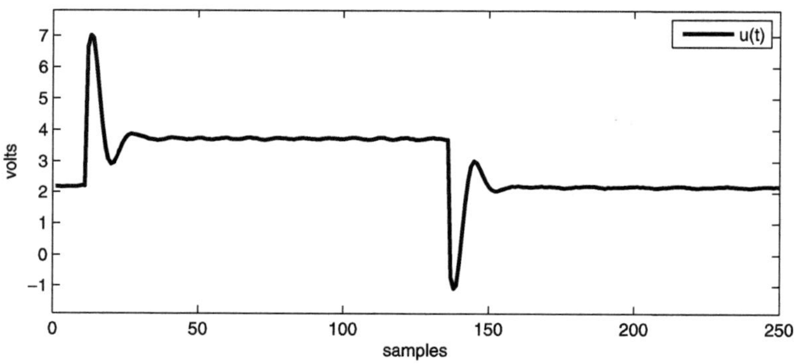

Fig. 8.14 Control signal of the PI controller obtained with the VRFT method for a reference model (8.7) with $a = 0.7$

Using the VRFT method with the instrumental variables estimate, the optimal parameter

$$\hat{\rho} = \begin{bmatrix} 0.1419 \\ -0.1287 \end{bmatrix}$$

is found, which corresponds to the following controller

$$C_{VR}(z, \hat{\rho}) = \frac{0.1419(z - 0.9069)}{z - 1}. \tag{8.8}$$

The closed-loop response $y(t)$ with this controller is presented in Fig. 8.13; the corresponding control signal $u(t)$ is presented in Fig. 8.14. It is seen that the closed-loop response is far from the desired one ($y_d(t)$), presenting an overshoot of 18% and a settling time significantly larger than specified. This visual perception can be quantified by calculating the resulting cost function: $\check{J}_T(\hat{\rho}) = 1.6033$ rad/s. Given that the reference's amplitude is 30 rad/s, one can say that the deviation from the

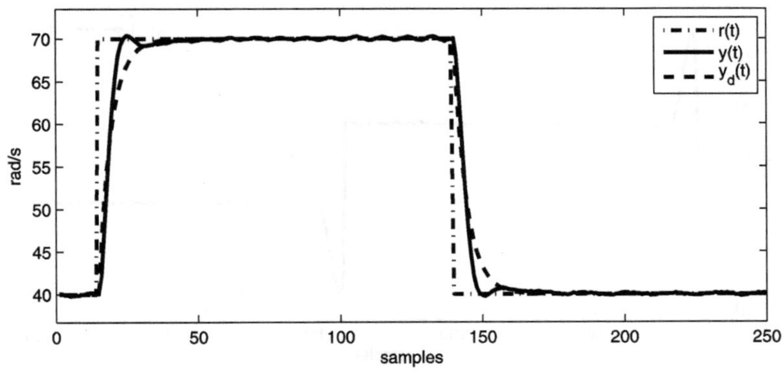

Fig. 8.15 Closed-loop responses of the reference model with $a = 0.8$ and the correspondent PI controller

desired output trajectory $y_d(t)$ is roughly 5.3% ($= 1.6/30$) in average for the whole measurement period, but this deviation is much larger during the transient.

This significant mismatch between the desired and the obtained closed-loop responses stems mainly from the significant violation of Assumption B_y. Note however that, despite this mismatch, the zero of the controller is close to the dominant pole, estimated from the open-loop response.

The choice of the parameter a was indeed not a very fortunate one. In view of the circumstances, a less ambitious performance should be specified. Still, the closed-loop obtained is stable, and its performance is better than what is usually found in control loops in industry.

But much better can be done with VRFT by choosing a reference model that is closer to the achievable performance. Let us perform a new direct design, now with $a = 0.8$. By using the same data as in the previous design, again with IV, the resulting controller is given by

$$C_{VR}(z, \hat{\rho}) = \frac{0.0951(z - 0.9011)}{z - 1}, \tag{8.9}$$

which corresponds to the optimal parameter

$$\hat{\rho} = \begin{bmatrix} 0.0951 \\ -0.08569 \end{bmatrix}.$$

Note that the zero of the controller's transfer function is very close to the one obtained in the first design, but now the controller's gain is smaller. The closed-loop response is presented in Fig. 8.15, and the cost is calculated as $\check{J}_T(\hat{\rho}) = 1.0116$ rad/s. The control signal $u(t)$ of this design is presented in Fig. 8.16. In this design, the overshoot has decreased significantly, though not completely vanished, and the settling time is not visually distinguishable from the desired one.

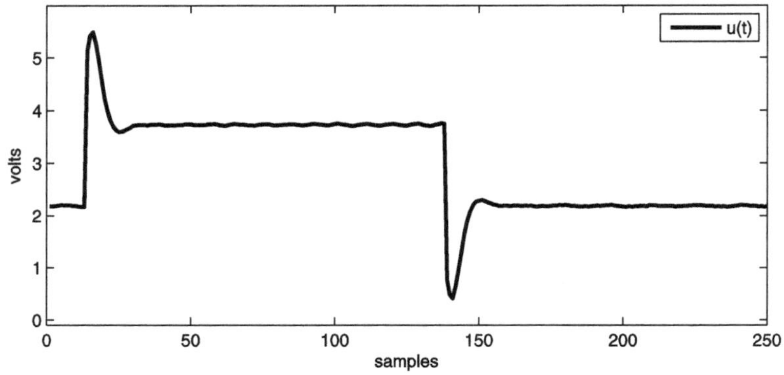

Fig. 8.16 Control signal of the PI controller obtained with the VRFT method for a reference model (8.7) with $a = 0.8$

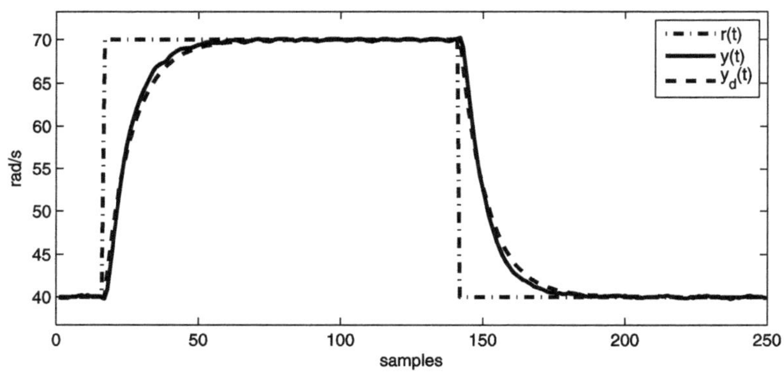

Fig. 8.17 Closed-loop responses of the reference model with $a = 0.9$ and the correspondent PI controller

Finally, let us use a reference model with the closed-loop pole at $a = 0.9$. The VRFT method estimates the controller

$$C_{VR}(z,\hat{\rho}) = \frac{0.0478(z - 0.8961)}{z - 1}, \tag{8.10}$$

which yields the closed-loop response presented in Fig. 8.17, resulting in a cost $\check{J}_T(\hat{\rho}) = 0.5604$ rad/s. This closed-loop response presents no overshoot and is very close to the desired one. The corresponding control signal is presented in Fig. 8.18. It is seen that as the parameter a is increased, that is, as the reference model's response becomes slower, the control signal decreases significantly in the transient period, as expected.

The results of the three designs that have been performed are summarized in Table 8.3. Notice that specifying a smaller settling time does not necessarily result is obtaining a smaller settling time. It does so only up to the point where the violation

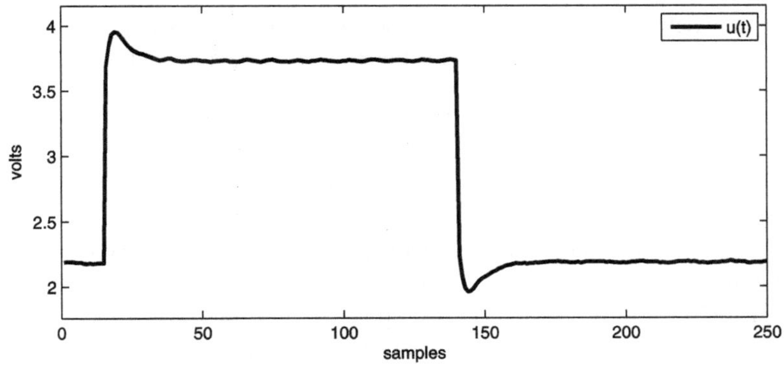

Fig. 8.18 Control signal of the PI controller obtained with the VRFT method for a reference model (8.7) with $a = 0.9$

Table 8.3 Different PI controllers obtained for different first order reference models (8.7) and the respective costs and settling times (specified and obtained)

a	num($C_{VR}(z, \hat{\rho})$)	$\check{J}_T(\hat{\rho})$ [rad/s]	t_s (spec)	t_s (obt)
0.7	$0.1419(z - 0.9069)$	1.6033	10	20
0.8	$0.0951(z - 0.9011)$	1.0116	15	15
0.9	$0.0478(z - 0.8961)$	0.5604	40	40

of Assumption B_y becomes too important. Also note that the PI controllers obtained with the VRFT method are such that their zeros almost cancels out the dominant pole of the plant, which is close to 0.9.

The results show that the VRFT method depends strongly on the matching condition, so it is highly desirable that Assumption B_y is verified or violated only moderately. Specifying a better performance (a faster reference model) does not necessarily result in obtaining a better performance! Specifying a performance that is not so good (a slower reference model), but is closer to the best that can be achieved with the controller class available, has yielded a much better performance in this practical application. Let us now explore the case where the reference model is chosen in order to (at least theoretically) respect the Assumption B_y.

8.2.1.2 Matched Case

In this section we will choose the reference model so that Assumption B_y is verified, under the hypothesis that the DC motor is described by a transfer function of order two as in (8.4). With the PI controller given in (8.3), whose transfer function can also be written as

$$C(z, \rho) = k_c \frac{z - f}{z - 1}, \tag{8.11}$$

with the new definition $k_c = k_p + k_i$ and the previously defined $f = \frac{k_p}{k_c}$, the closed-loop system is described by the transfer function (8.5), reproduced here for ease of reference

$$T(z,\rho)\frac{k_g k_c z(z-f)}{(z-1)(z-b)(z-c)+k_g k_c z(z-f)}.$$

It suffices to choose a reference model in this form to satisfy Assumption B_y. To make our lives easier, let us pick a second order reference model, which corresponds to the cancellation of one of the process' poles by the controller's zero; let us say $f = c$. Then the closed-loop transfer function is in the form

$$T(z,\rho) = \frac{k_g k_c z}{(z-1)(z-b)+k_g k_c z} = \frac{k_g k_c z}{z^2+(k_g k_c - b - 1)z + b}. \tag{8.12}$$

The model reference $T_d(z)$ must be chosen such that (8.12) equals $T_d(z)$ for some value of k_c. This will happen for any real number K in the following transfer function

$$T_d(z) = \frac{Kz}{z^2+(K-b-1)z+b}.$$

Notice that choosing this reference model requires knowledge of one of the process' poles—the parameter b. As expected, enforcing Assumption B_y requires some additional information about the process. But a reasonable estimate for the dominant pole is already available from the open-loop step response presented previously. So, we can easily define a second order reference model (8.12), and the matching condition is satisfied. Finally, we still have to choose the parameter K to completely define the reference model $T_d(z)$. Any value of K will do, but of course for each different value a different performance is specified.

Let us show the results obtained with $b = 0.9$ and $K = 0.005$, which substituted into (8.12) yields

$$T_d(z) = \frac{0.005z}{z^2 - 1.895z + 0.9}.$$

Then the controller obtained using the VRFT method is given by

$$C_{VR}(z,\hat{\rho}) = \frac{0.0020(z+0.2585)}{z-1} \tag{8.13}$$

which resulted in a cost of $\check{J}_T(\hat{\rho}) = 0.3424$ rad/s. Notice that this cost is lower compared to all the costs presented in Table 8.3, and results almost exclusively from the effect of noise, that is, $J_y(\hat{\rho}) \approx 0$. The closed-loop response obtained using this controller, compared to the desired closed-loop response, is presented in Fig. 8.19. Now that Assumption B_y is verified, the closed-loop response is very close to the response of the model reference. The control signal is presented in Fig. 8.20.

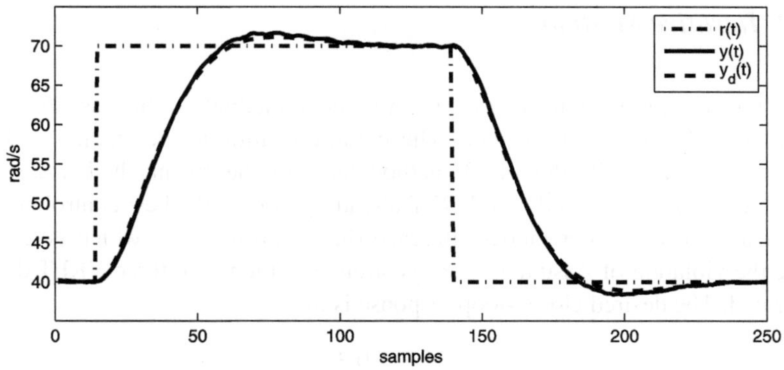

Fig. 8.19 Closed-loop responses of the reference model (8.12) with $a = 0.9$ and $k = 0.005$ and the correspondent PI controller

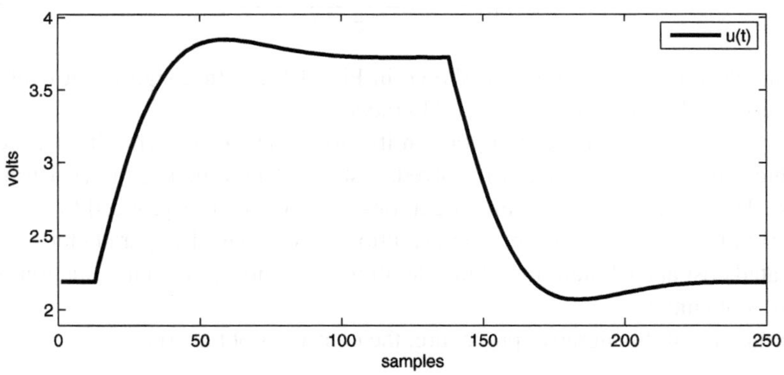

Fig. 8.20 Control signal of the PI controller obtained with the VRFT method for a reference model (8.7) with $a = 0.9$ and $k = 0.005$

One could try other values of K, but this will not result in a reference model with better transient response. Indeed, by changing the value of K the poles of the reference model (and hence of the closed-loop system) change, but their product is always equal to b—the dominant pole of the process. Since the poles of the closed loop must always have module smaller than one, this implies that both poles have module larger than b. As a consequence, it is not possible to obtain a closed-loop response faster than the open-loop response. It would be possible in principle to assign the other pole of the process to b, instead of assigning the dominant pole, which would result in faster reference models and hence faster closed-loop performance. But this would require knowing the value of this other pole at least approximately, an information that is not available to the designer.

8.2.2 Iterative Method

Let us now design the controller using an iterative method, in this case the Correlation based Tuning (CbT) method. The initial condition to the method will be a controller designed with the VRFT method and only the mismatched case will be considered—in the matched case VRFT already produces the best controller. Consider again the first-order model reference (8.7) with $a = 0.7$, which is the case where the violation of Assumption B_y is strongest among the three VRFT designs performed. The desired closed-loop response is then

$$T_d(z) = \frac{0.3}{z - 0.7}. \tag{8.14}$$

The VRFT design has yielded the following controller

$$C(z, \rho_0) = \frac{0.1419(z - 0.9069)}{z - 1}, \tag{8.15}$$

and the closed-loop performance shown in Fig. 8.13, with a significant overshoot and a cost evaluated at $\check{J}_T(\rho_0) = 1.6033$ rad/s.

The CbT method has been applied in its most convenient form, that is, the one in which the correlation (7.25) is solved, instead of minimizing the cost function (7.28). The algorithm ran for three iterations and the cost was reduced to $\check{J}_T(\rho_3) = 1.2070$ rad/s. Table 8.4 presents the evolution of the controller parameters and the estimated cost at each iteration. After iteration three, no significant reduction of the cost was obtained.

At the end of this iterative procedure, the controller obtained is

$$C(z, \rho_3) = \frac{0.1242z - 0.1142}{z - 1}, \tag{8.16}$$

which yields the closed-loop output presented in Fig. 8.21. The overshoot has been decreased from 16% in the initial controller to 6%.

The iterative procedure gets much closer to the minimum of $J_y(\rho)$ than the VRFT design. This results in a significant performance improvement regarding this performance criterion. Yet, the controller design still suffers from the strong violation of the matching condition.

Table 8.4 Controller evolution with CbT

i	$\mathrm{num}(C_{Cb}(z, \rho_i))$	$\check{J}_T(\rho_i)$ [rad/s]
0	$0.1419z - 0.1286$	1.6033
1	$0.1303z - 0.1198$	1.3190
2	$0.1238z - 0.1137$	1.2843
3	$0.1242z - 0.1142$	1.2070

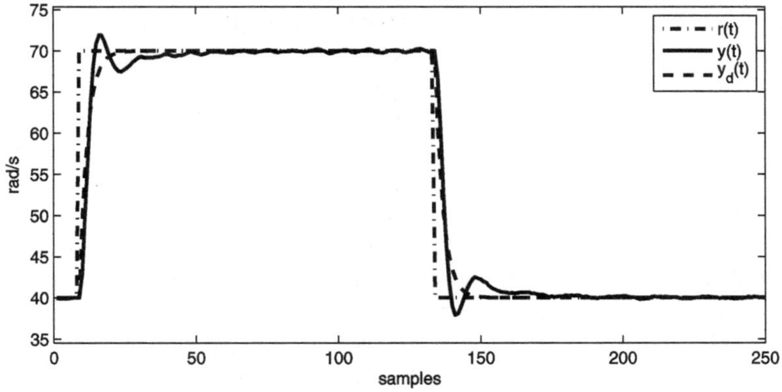

Fig. 8.21 Closed-loop responses of the reference model with $a = 0.7$ and the correspondent PI controller $C(z, \rho_3)$ obtained with CbT

8.3 A Temperature Process

The third process to be studied consists of a resistance whose temperature is measured by a thermopair and controlled by a commercial PID controller. The output of the process is the temperature, which is given in degrees Celsius. The input of the process is the power applied to the resistance and is expressed as a percentage of the maximum power that can be applied to the system. The sampling period is 3 seconds.

Results in two different scenarios will be presented. In the first scenario, the signal-to-noise (SNR) ratio is comfortably large and Assumption B_y is satisfied. This scenario resembles the ideal conditions for application of direct data-driven design (VRFT) and is accordingly called the "favorable scenario". For a second scenario the SNR will be decreased significantly and Assumption B_y will not be satisfied. The process is the same, and so is its noise level. The SNR is changed by reducing the amplitude of the reference, and Assumption B_y will no longer be satisfied because a simpler class of controllers will be used: a PI controller. In this situation the data-driven design becomes more involved; accordingly this will be called the "tough scenario".

8.3.1 The Favorable Scenario

In this scenario, a PID controller with the following structure

$$C(z, \rho) = [k_p \; k_i \; k_d] \begin{bmatrix} 1 \\ \frac{z}{z-1} \\ \frac{z-1}{z} \end{bmatrix},$$

will be designed. The behavior desired for the closed loop is defined by the reference model

$$T_d(z) = \frac{0.1}{z - 0.9}.$$

With this reference model and model class, Assumption B_y is satisfied provided that the process' transfer function is of order at most equal to two. The literature on modeling of electrically heated temperature processes indicates that this is the case here.[4] The process operates at temperatures around 80°C and the references to be followed are step changes of amplitude equal to 20°C. Given that the system noise is under 1°C, this provides a large SNR. In the experiments performed for data collection the reference signal will have a similar amplitude, so that the signal-to-noise ratio is high and the system noise will have little effect on the controller design. This large SNR and full order controller (Assumption B_y satisfied) is the ideal situation from a control design perspective, which is the reason why it is called the "favorable scenario". We will design the controller using a direct and an iterative method, starting with the direct method.

8.3.1.1 Direct Method

The Virtual Reference Feedback Tuning with instrumental variables is used to optimize the reference tracking performance. Two open-loop experiments are performed to collect the data and form the instrumental variable to be used in the VRFT calculations. The same input was used in both experiments:

$$u(t) = 10 + 5\sin\left(\frac{2\pi t}{60}\right) - 5\sin\left(\frac{2\pi t}{30}\right) + 5\sin\left(\frac{2\pi t}{10}\right)$$

and $N = 420$ samples of this signal (1260 seconds) have been collected. The input signal ($u(t)$) is shown in Fig. 8.22 and the resulting output signals of each experiment ($y_1(t)$ and $y_2(t)$) are shown in Fig. 8.23. Notice that this input is sufficiently rich of order seven (SR7), much more than the minimum richness required for optimization of the three parameters in the PID controller.

The data collected in the open-loop experiment were utilized to design a PID controller with the VRFT-IV method, with the IV generated by the repeated experiment. The following controller has been obtained:

$$C_{VR}(z, \hat{\rho}) = \begin{bmatrix} 0.6684 & 0.0178 & 0.6808 \end{bmatrix} \begin{bmatrix} 1 \\ \frac{z}{z-1} \\ \frac{z-1}{z} \end{bmatrix}. \tag{8.17}$$

[4] We will show later, in Sect. 8.3.1.3 that this is indeed the case.

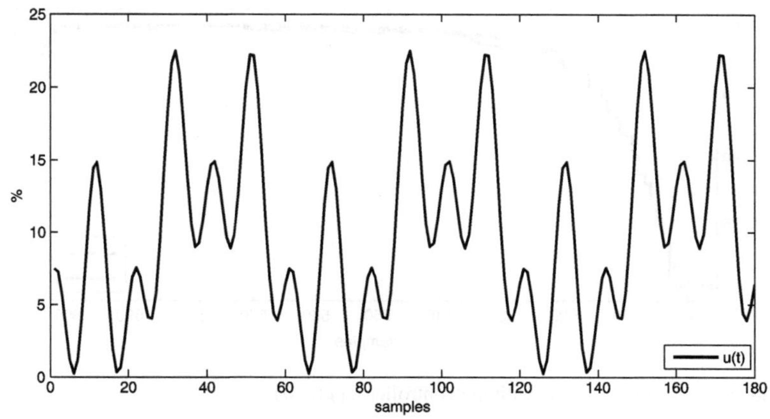

Fig. 8.22 Input collected from the open-loop experiment

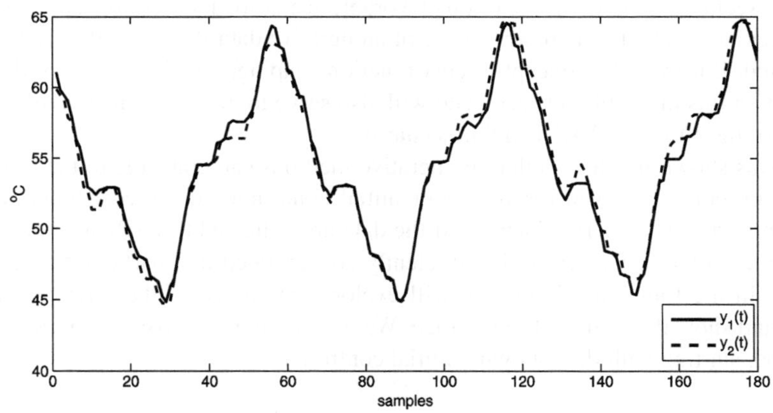

Fig. 8.23 Output collected from the open-loop experiment

The controller $C_{VR}(z, \hat{\rho})$ just designed was included in the loop and a reference signal consisting of a step change from 70°C to 90°C was applied. Figure 8.24 shows the resulting closed-loop response $y(t, \hat{\rho})$, along with the desired response $y_d(t)$ (the response of the reference model to this same reference signal). As expected from the favorable conditions in this scenario, the resulting response is very similar to the desired one, showing that the VRFT method is indicated for this case. It can be inferred from the plots that the difference between the two signals is due exclusively, or almost, to the noise; in other words, $J_y(\hat{\rho}) \approx 0$ and thus $J_T(\hat{\rho}) \approx J_e(\hat{\rho})$. The cost is estimated with 100 samples of the signals plotted in Fig. 8.24, resulting in $\check{J}(\hat{\rho}) = 0.26$°C per sample. Since Assumption B_y is satisfied and the SNR is large, it is expected that $C_{VR}(z, \hat{\rho})$, obtained through the application of the VRFT method, will be close to $C_d(z)$.

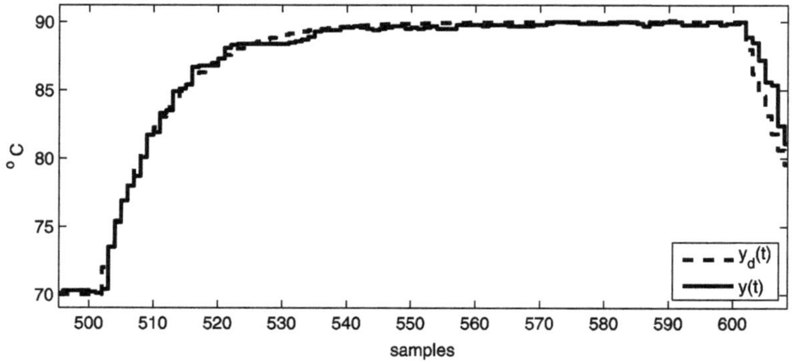

Fig. 8.24 Closed-loop response with the controller $C_{VR}(z, \hat{\rho})$

8.3.1.2 Iterative Method

VRFT seems to fit like a glove to this favorable scenario, and iterative optimization unnecessary. But let us explore the use of an iterative data-driven method anyways. This will illustrate the concept of cost function shaping, which are observed more clearly in this favorable scenario, and will also serve to give even more evidence in favor of the choice of VRFT in this scenario.

It was shown in Chap. 6 that the iterative methods can converge to the globally optimum controller even for very poor initial conditions, if the cost function has an appropriate shape. This happens if the data are collected through an experiment where the reference spectrum is sufficiently concentrated in a particular frequency range. In the following design we will explore the choice of the reference spectrum and show the results of this choice. We will use the Iterative Feedback Tuning method, starting with the following initial controller

$$C(z, \rho_0) = [0.55 \;\; 0.02 \;\; 0.00] \begin{bmatrix} 1 \\ \frac{z}{z-1} \\ \frac{z-1}{z} \end{bmatrix}, \tag{8.18}$$

which is far from optimal. Figure 8.25 shows the step response from 70°C to 90°C with the controller $C(z, \rho_0)$. Using 100 samples of this signal the cost function is evaluated to $\check{J}_T(\rho_0) = 6.34$°C per sample.

Initially, the following reference signal is applied:

$$r_1(t) = 80 + 10sq\left(\frac{2\pi t}{200}\right).$$

This reference signal, which is not the reference that the system is supposed to track, is applied for controller tuning purposes, in order to guarantee persistence of excitation. This signal is sufficiently rich of order three (even if only the first harmonic is considered), as required, and causes the process to operate around 80°C; its amplitude is 20°C, guaranteeing a large SNR. It is likely that the application

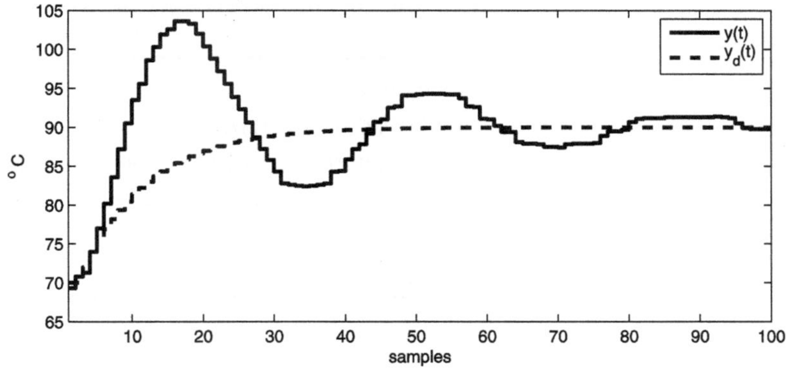

Fig. 8.25 Closed-loop response with the initial controller $C(z, \rho_0)$

of such a large probe signal would not be viable in many practical industrial applications, which is one reason why the conditions are not usually as favorable in practice. The spectrum of this reference is concentrated at a very low frequency range, which is appropriate to eliminate the possibility of local maxima or minima in the search space. That is, by choosing this particular reference the cost function has been shaped to have no other extrema than the global minimum, as explained in Chap. 6.

The IFT method actually minimizes the composite cost $J_T(\rho)$, but given the large SNR, $J_T(\rho) \approx J_y(\rho)$. On the other hand, under Assumption B_y the global minimum of $J_y(\rho)$ is always the same regardless of the reference input. So, the global minimum in the following optimization will not be significantly changed by changing the reference from a step to the reference $r_1(t)$ above during the tuning procedure.

The IFT algorithm ran for five iterations using the step size sequence proposed by the first method presented in the Chap. 5, Sect. 5.4.2. The controller parameters at each iteration are shown in Table 8.5, along with the cost evaluated at each iteration. After five iterations the cost was reduced to $\check{J}_T(\rho_5) = 0.495°\text{C}$, achieving the following controller:

$$C(z, \rho_5) = [0.7168 \;\; 0.01954 \;\; -0.053] \begin{bmatrix} 1 \\ \frac{z}{z-1} \\ \frac{z-1}{z} \end{bmatrix}.$$

We observe in Table 8.5 the convergence of the controller parameter; convergence after iteration number five becomes very slow and is not presented. The parameters k_p and k_i obtained using the IFT method (in $C(z, \rho_5)$) are very close to their corresponding values in the controller designed by VRFT—$C_{VR}(z, \hat{\rho})$. The derivative gain k_d, on the other hand, is smaller; this is to be expected since the derivative action tends to amplify the noise and this effect is penalized in the performance criterion $J_T(\rho)$, but not in $J_y(\rho)$. The step response of the closed-loop system with $C(z, \rho_5)$ is shown in Fig. 8.26, where it is seen that the performance is

Table 8.5 Controller evolution $C(z, \rho_i)$ using the reference signal $r_1(t)$

i	ρ_i^T	μ_i	$\check{J}_T^1(\rho_i)$
0	[0.5500 0.20000 0.00000]	0.0930	6.34
1	[0.6973 0.03970 −0.04672]	0.0116	1.76
2	[0.6641 0.01956 −0.05345]	0.0058	0.67
3	[0.7069 0.02960 −0.04977]	0.0003	1.23
4	[0.7118 0.02457 −0.05141]	0.0003	0.88
5	[0.7168 0.01954 −0.05304]	0.0003	0.49

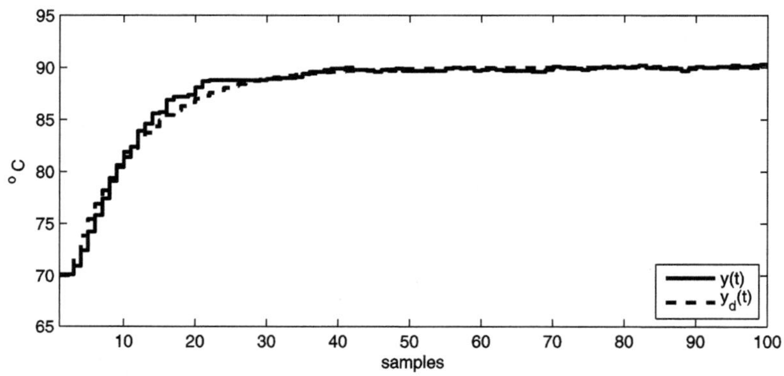

Fig. 8.26 Closed-loop response with the controller $C(z, \rho_5)$

close to, though noticeably different from, the one obtained with the VRFT method, presented in Fig. 8.24.

Let us now explore the choice of the reference signal used to compute $\widehat{\nabla J}_T(\rho)$. This choice is critical to ensure convergence of the iterative algorithm to the global minimum because the shape of the cost function is strongly related to the reference signal, as seen in Chap. 6.

Consider the following reference signal

$$r_2(t) = 80 + 10 \sin\left(\frac{2\pi t}{58}\right)$$

whose spectrum spreads to higher frequencies than the spectrum of $r_1(t)$. Applying the IFT method with the same initial controller $C(z, \rho_0)$, but now with data collected from an experiment using $r_2(t)$ as the reference signal, the evolution of the controller parameter is as shown in Table 8.6. Observe that the cost function being minimized is not the same as before, because the reference signals are different. Accordingly, we have named this new cost function $J_T^2(\rho)$, in opposition to the notation $J_T^1(\rho)$ used for the cost function resulting from the reference $r_1(t)$.

Table 8.6 Controller evolution $C(z, \rho_i)$ using the reference signal $r_2(t)$

i	ρ_i^T	μ_i	$\check{J}_T^2(\rho_i)$
0	[0.5500 0.2000 0.0000]	0.0116	3.60
1	[0.5327 0.2343 0.0029]	0.0116	3.51
2	[0.4996 0.2738 0.0128]	0.0116	3.43
3	[0.4730 0.3116 0.0197]	0.0116	3.39

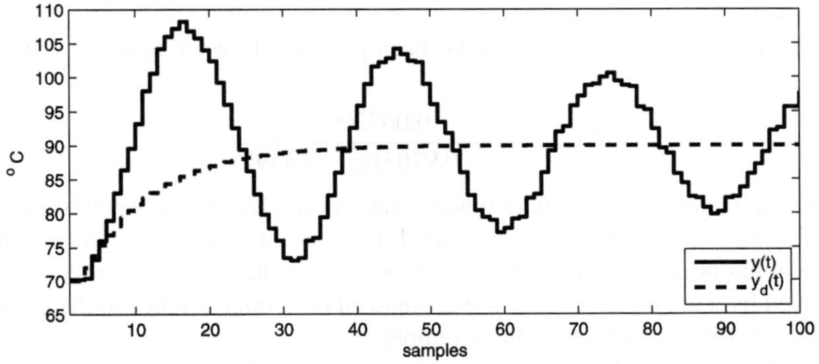

Fig. 8.27 Closed-loop response with the controller $C(z, \rho_3)$

After three iterations, the controller

$$C(z, \rho_3) = [0.4730\ \ 0.3116\ \ 0.0197] \begin{bmatrix} 1 \\ \frac{z}{z-1} \\ \frac{z-1}{z} \end{bmatrix}$$

has been achieved and the cost has been reduced from $\check{J}_T^2(\rho_0) = 3.60°\text{C}$ to $\check{J}_T^2(\rho_3) = 3.39°\text{C}$ per sample. Additional iterations result in very little change in the controller parameters and the cost function's value. The behavior of the closed-loop system with the controller $C(z, \rho_3)$ is shown in Fig. 8.27.

Despite the fact that the cost is being reduced, the parameters of the controller are getting farther away form the global minimum; the algorithm seems to be converging to a local minimum of $J_T^2(\rho)$. The initial controller $C(z, \rho_0)$ is not in the domain of attraction of the global minimum. The estimated cost calculated with a step reference is $\check{J}_T^{step}(\rho_3) = 10.00°\text{C}$, significantly higher than the $\check{J}_T^{step}(\rho_0) = 6.34°\text{C}$ obtained with the initial controller $C(z, \rho_0)$ for the same step reference.

8.3.1.3 A Performance Benchmark—Model-Based Design

In the following, a model-based design for this temperature process will be briefly presented. The purpose is to establish a benchmark for the performance that can be

Table 8.7 Comparison between the temperature controllers obtained using different methods

Method	Controller	Optimal ρ^T	$\check{J}_T(\rho)$
model-based	$C(z, \rho_d)$	[0.6750 0.0200 0.7930]	0.22
VRFT	$C_{VR}(z, \rho)$	[0.6684 0.0178 0.6808]	0.26
IFT	$C(z, \rho_5)$	[0.7168 0.0195 −0.0530]	0.49

achieved, and thus assess the quality of the results obtained with the VRFT and IFT methods.

A prediction error identification has been performed, resulting in the following model:

$$G(z) = \frac{0.067209z}{(z - 0.9703)(z - 0.5492)}.$$

The parameters of this transfer function have been identified to a precision better than 3×10^{-3}. This model was not used at all in the control designs presented before; it has been identified only to confirm the hypothesis that the process is well described by a model of order two and the ideal controller belongs to the class we considered (PID controllers) in this example.

Based on this model, the ideal controller is calculated as

$$C_d(z) = [0.675 \ \ 0.02 \ \ 0.793] \begin{bmatrix} 1 \\ \frac{z}{z-1} \\ \frac{z-1}{z} \end{bmatrix},$$

whose parameters are similar to the ones obtained using the VRFT method. Again, k_p and k_i obtained using the IFT method are still similar to ρ_d, as shown before.

Applying a reference step to the closed-loop system with the controller $C_d(z)$ results in $\check{J}_T(\rho_d) = 0.22$°C. Notice that $J_y(\rho_d) \cong 0$, which implies that the cost obtained is due only to the noise contribution; that is, it is fair to say that

$$\check{J}_T(\rho_d) \cong \check{J}_e(\rho_d) \cong 0.22\text{°C}.$$

This is thus the optimal value that could be expected for $J_T(\rho)$ when using VRFT, and it is also indicative of the value that we can expect to obtain with IFT. The results obtained in each method are compared in Table 8.7.

8.3.2 Tough Case

Consider now the control of the same temperature process, with the same reference model

$$T_d(z) = \frac{1-a}{z-a},$$

but now with the controller constrained to a PI structure

$$C(z,\rho) = [\varrho_1\ \varrho_2] \begin{bmatrix} \frac{z}{z-1} \\ \frac{1}{z-1} \end{bmatrix}.$$

With this controller class and reference model, Assumption B_y would require the process model to be of order one; we are now in the mismatched case. Besides, the design will be carried out with references of amplitude equal to five, so that the SNR is significantly smaller than in the previous scenario. These two aspects complicate the data-driven design, which is why this is called the "tough scenario".

8.3.2.1 Direct Method

The VRFT method will be applied using both the IV and LS estimates; this exercise will serve the purpose of illustrating the statistical properties of both estimates, quantifying them in a practical example. We will design the controllers for three different values of the parameter a in the reference model: $a = 0.8$, $a = 0.9$ and $a = 0.95$.

In this mismatched control situation, the signals must be filtered in order to approximate the minima of $J^{VR}(\rho)$ and $J_y(\rho)$. The input signal will be a square wave applied in open-loop, so that the filter can be computed as in (3.18), where the spectrum of the reference signal and the spectrum of the input signal are the same. So, the filter's transfer function is given by

$$L(z) = T_d(z)(1 - T_d(z)).$$

The controller parameter resulting from VRFT design is a random variable whose statistical properties we want to assess. In order to do that, 100 Monte Carlo experiments have been performed to collect data. The input for each one of these experiments was the same, as it is shown in Fig. 8.28; the output for one of the experiments is presented in Fig. 8.29. It is seen that the noise level is far from negligible.

Since data from two experiments are needed to design a controller using the instrumental variables approach, these 100 data sets were combined in pairs, resulting in 50 pairs. For each pair of data sets, a controller has been designed by means of VRFT with the IV approach. We have also designed 50 different controllers using the least squares estimate, in each case using one of the data sets of the corresponding pair. Each one of these 100 controllers designed by VRFT—50 designed with IV and 50 designed with LS—was applied to the system. This whole procedure has been repeated for the three reference models—$a = 0.8$, $a = 0.9$ and $a = 0.95$. The results obtained for all these 300 controllers are presented and analyzed next.

First of all, let us verify the average performance obtained with each method and each reference model. To do that, calculate the mean controller in each case, which

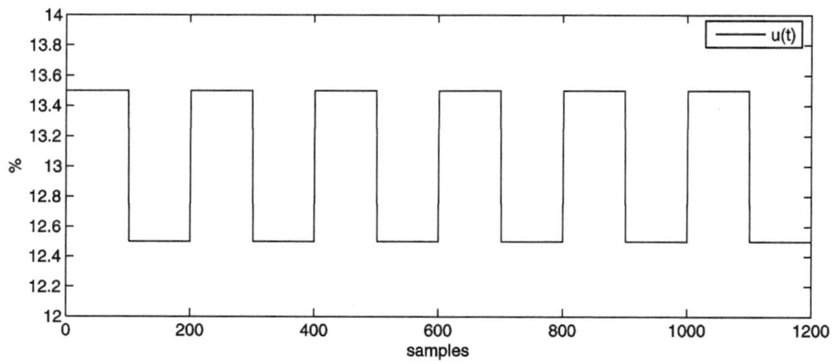

Fig. 8.28 Input of one open-loop experiment in the thermal process

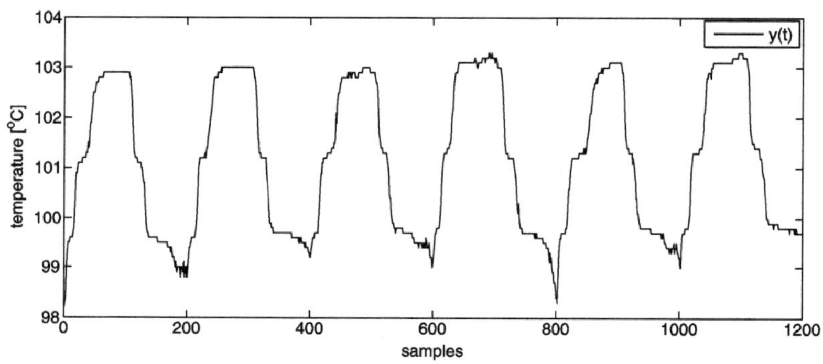

Fig. 8.29 Output of one open-loop experiment in the thermal process

is given by

$$\rho_m = \frac{1}{50}\sum_{i=1}^{50} \rho_i.$$

Table 8.8 shows all these mean controllers and the corresponding costs, which are considerably lower with the IV estimate, as expected. Notice that the controllers obtained with the IV estimates approximately cancel out the process' dominant pole at $z = 0.97$, just like the ideal controller $C_d(z)$ calculated in Sect. 8.3.1.3 did. It is also worthwhile to compare these mean controllers to $C_d(z)$, evaluating the mismatch factor $K(z, \rho_m) = C(z, \rho_m) - C_d(z)$.

The step responses of the closed-loop system with the mean controllers are presented in Figs. 8.30–8.35. Figures 8.30, 8.31, 8.32 and 8.33 show the step response and the controller's output for $a = 0.8$, obtained using LS and IV (the mean controller in each case). The step response obtained using IV is significantly closer to the desired response than the one obtained with LS. Figures 8.34 and 8.35 present the step responses (we omit the control signal in this case) obtained when using $a = 0.95$. The graphic results confirm the results presented in Table 8.8: for this

Table 8.8 Mean value of the controllers for different values of a

a	LS		IV	
	$\text{num}(C_{LS}(z, \rho_m))$	$\check{J}_T(\rho)$	$\text{num}(C_{IV}(z, \rho_m))$	$\check{J}_T(\rho)$
0.80	$0.27590(z - 0.8986)$	4.1296	$1.34440(z - 0.9717)$	1.4012
0.90	$0.33838(z - 0.9474)$	2.1279	$0.70074(z - 0.9709)$	0.5869
0.95	$0.27462(z - 0.9648)$	0.7479	$0.33938(z - 0.9704)$	0.4696

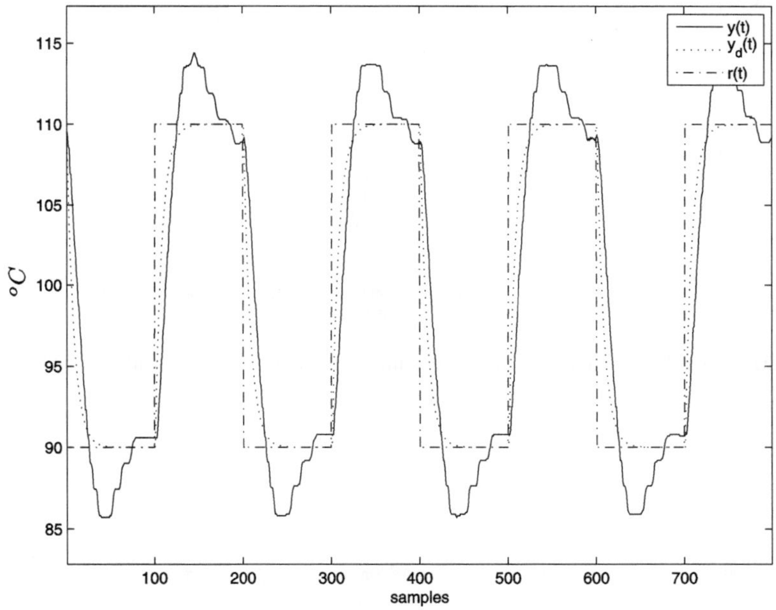

Fig. 8.30 Step response with the mean controller obtained for $T_d(z)$ with $a = 0.80$ and LS

choice of reference model, the IV estimate still gives a better result, but the difference is much smaller than for $a = 0.8$.

Let us now analyze the distribution of the controllers around the mean value, by computing the sample variance of the controller parameters. Figure 8.36 shows the parameters of the 100 controllers designed based on the reference model with $a = 0.90$. The following data are provided for each estimate (IV and LS): the 50 parameters obtained, the mean value of the parameters, and an ellipse that represents the covariance. Specifically, each ellipse represents the corresponding 95% confidence interval, that is, each ellipse encloses 95% of the parameters calculated. It is observed that the variance of the parameters computed with IV is considerably larger than the variance of the parameters computed with LS, but it has also been seen that the LS yields in average a worse performance—a more biased estimate of $C_d(z)$. Figure 8.37 shows the same results for the reference model with $a = 0.95$.

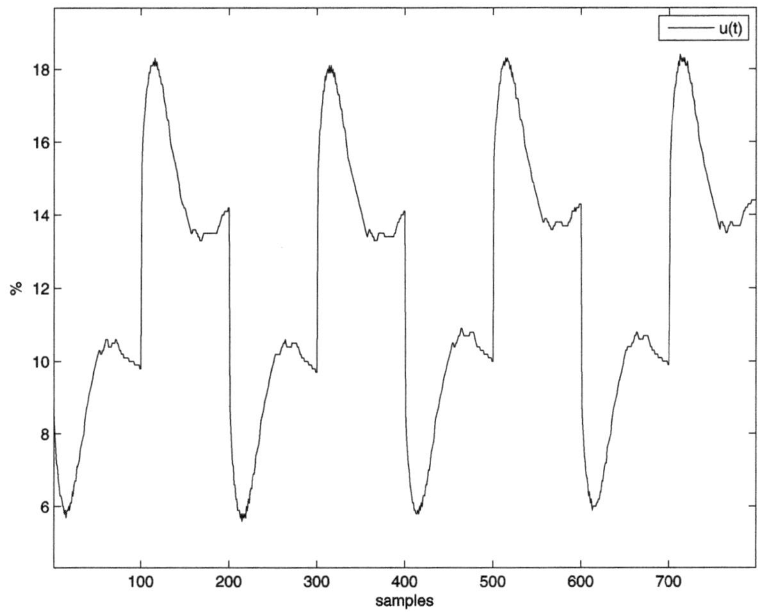

Fig. 8.31 Controller's output with the mean controller obtained for $T_d(z)$ with $a = 0.80$ and LS

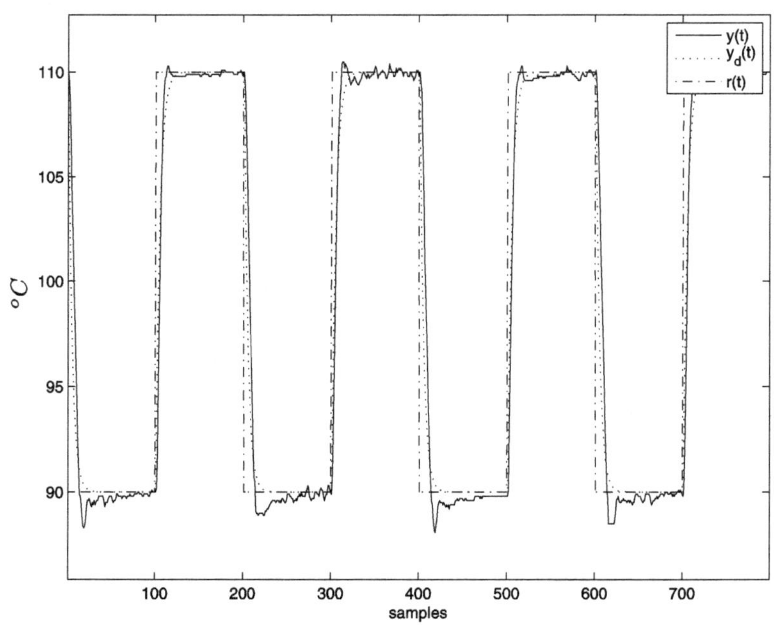

Fig. 8.32 Step response with the mean controller obtained for $T_d(z)$ with $a = 0.80$ and IV

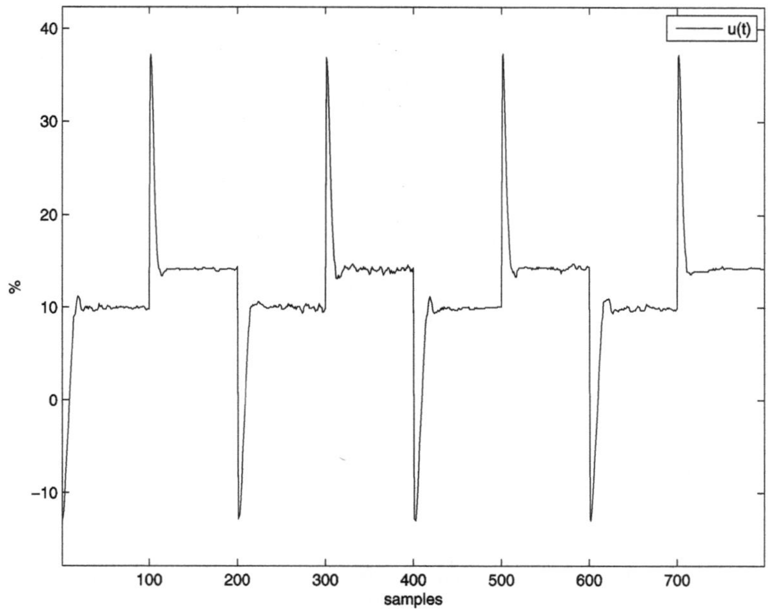

Fig. 8.33 Controller's output with the mean controller obtained for $T_d(z)$ with $a = 0.80$ and IV. The controller's output goes below zero, but this is not applied to the process—the control action is saturated at zero

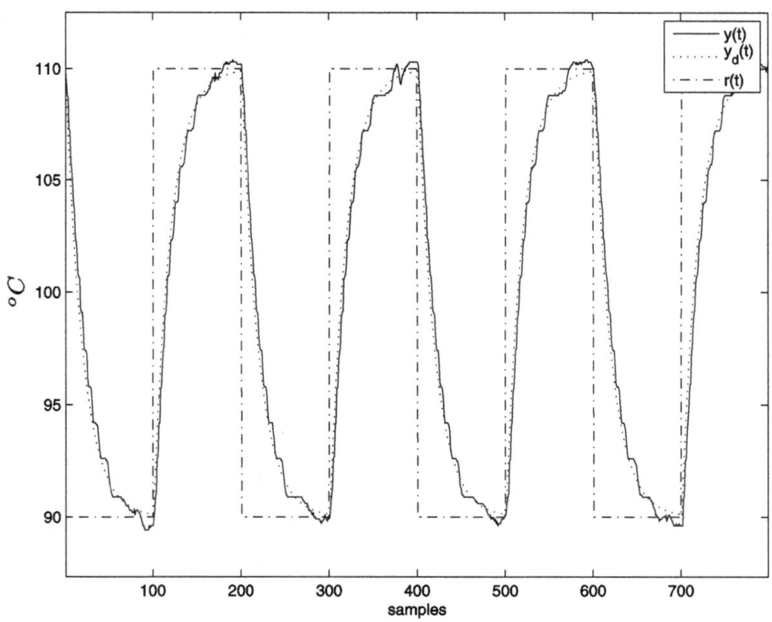

Fig. 8.34 Step response with the mean controller obtained for $T_d(z)$ with $a = 0.95$ and LS

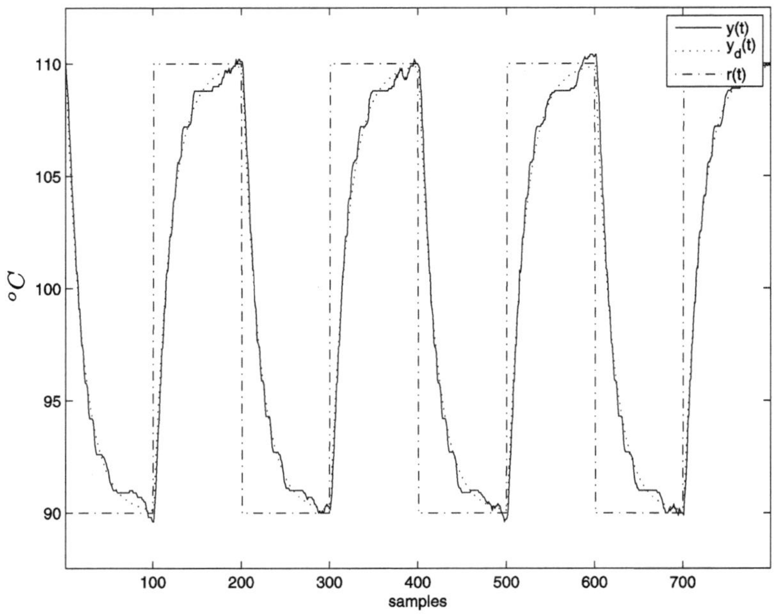

Fig. 8.35 Step response with the mean controller obtained for $T_d(z)$ with $a = 0.95$ and IV

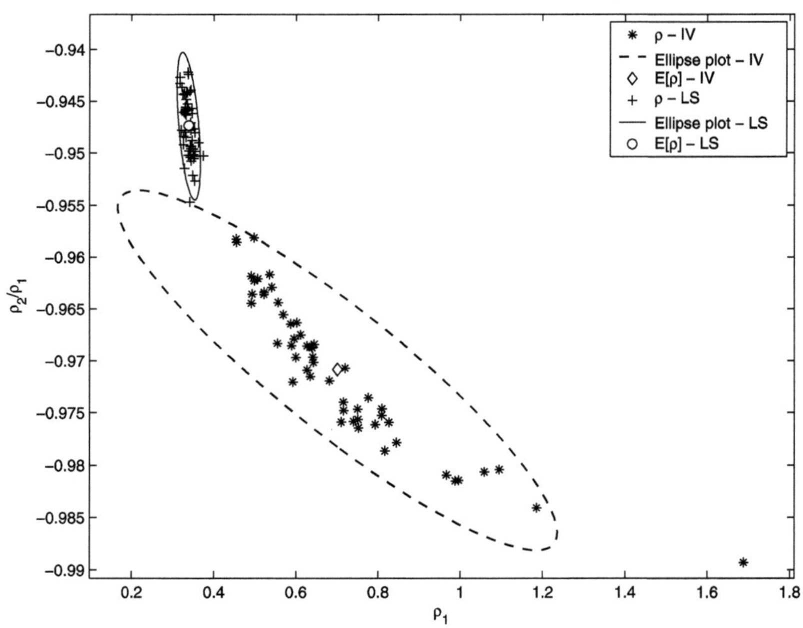

Fig. 8.36 Ellipse plots (using IV and LS) estimated from 50 parameters obtained for $T_d(z)$ with $a = 0.90$

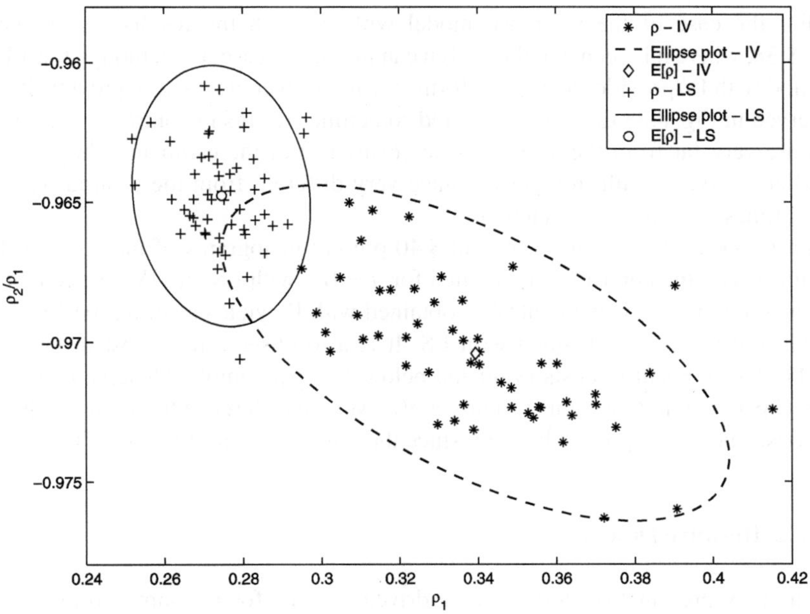

Fig. 8.37 Ellipse plots (using IV and LS) estimated from 50 parameters obtained for $T_d(z)$ with $a = 0.95$

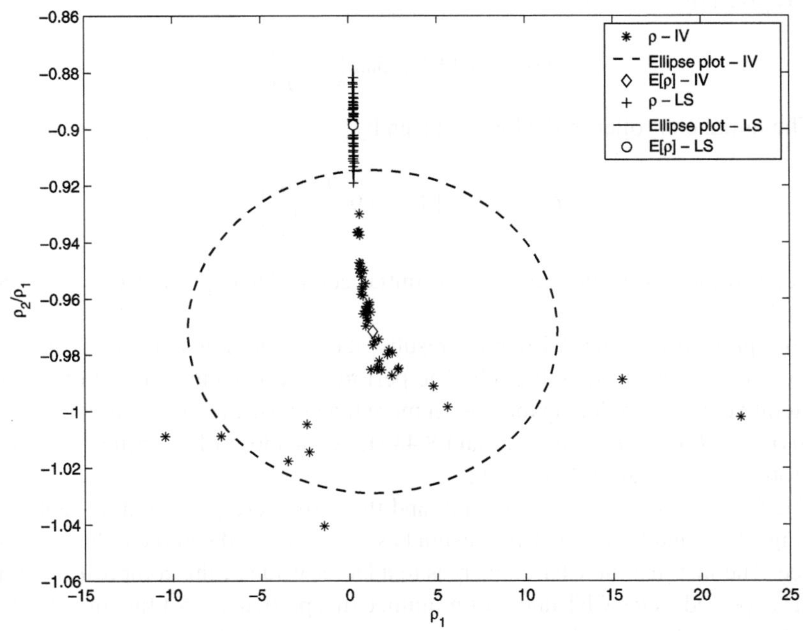

Fig. 8.38 Ellipse plots (using IV and LS) estimated from 50 parameters obtained for $T_d(z)$ with $a = 0.80$

For the case of the reference model with $a = 0.8$ the results are shown in Fig. 8.38, where it is seen that the design can fail in this case. Even though the VRFT method with IV provides a good performance in the average, as seen previously, it is observed in Fig. 8.38 that the IV method sometimes results in controller parameters that are very far from the average—the covariance of the estimate is huge. These "outliers" would result in a performance very different from the desired one, and many times in an unstable behavior.

Last, but not least, Figs. 8.39 and 8.40 present histograms of the cost function obtained with the controllers computed for $a = 0.9$ with LS and IV, respectively. It can be seen that the worst controller obtained with IV yields a similar performance to the best controller obtained with LS. It is also observe that most instances of the IV design result in costs evaluated below 1 °C per sample. Observe that a cost of 1 °C per sample represents roughly a 5% average difference between the desired response and the response obtained, since the reference amplitude is 20°C.

8.3.2.2 Iterative Design

Let us now present two iterative data-driven designs for the same problem. One design will be made to optimize reference tracking performance, using the CbT method, and another will minimize the composite performance criterion, using the IFT method; both use the first-order reference model with $a = 0.9$. The iterative methods use data collected in closed loop, and the reference signal used to collect data is given by

$$r(t) = 100 + 2.5sq\left(\frac{2\pi}{200}\right).$$

The initial controller in the loop is given by

$$C(z, \rho_0) = [1 \;\; -0.9]\begin{bmatrix} \frac{z}{z-1} \\ \frac{1}{z-1} \end{bmatrix}.$$

The closed-loop performance with the initial controller is presented in Figs. 8.41 and 8.42.

The application of the IFT method results, at each iteration, in the controllers and the cost values shown in Table 8.9. The performance is significantly improved by the iterative design, reducing the cost in more than one order of magnitude, down to $\check{J}_T(\rho_9) = 0.065875$. Figures 8.43 and 8.44 show the closed-loop behavior with the controller obtained after 9 iterations.

The CbT method was also applied, and the results are presented in Table 8.10, and Figs. 8.45 and 8.46. The CbT design has also improved significantly the performance. The fact that the CbT converges to a larger value of the composite cost than IFT is expected, since CbT does not minimize this particular cost function. Still, the performances obtained with each one of the two iterative design methods are quite similar.

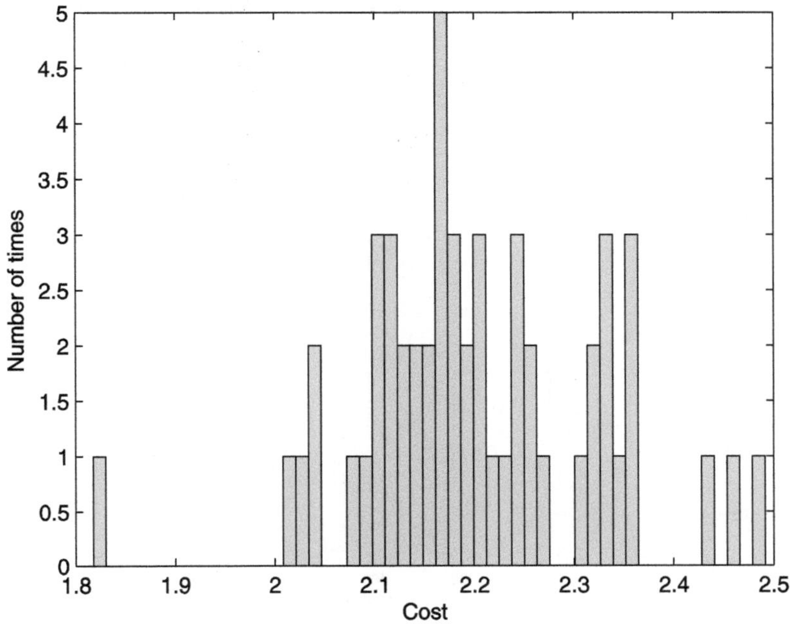

Fig. 8.39 Histogram plot obtained for a $T_d(z)$ with $a = 0.90$ using LS

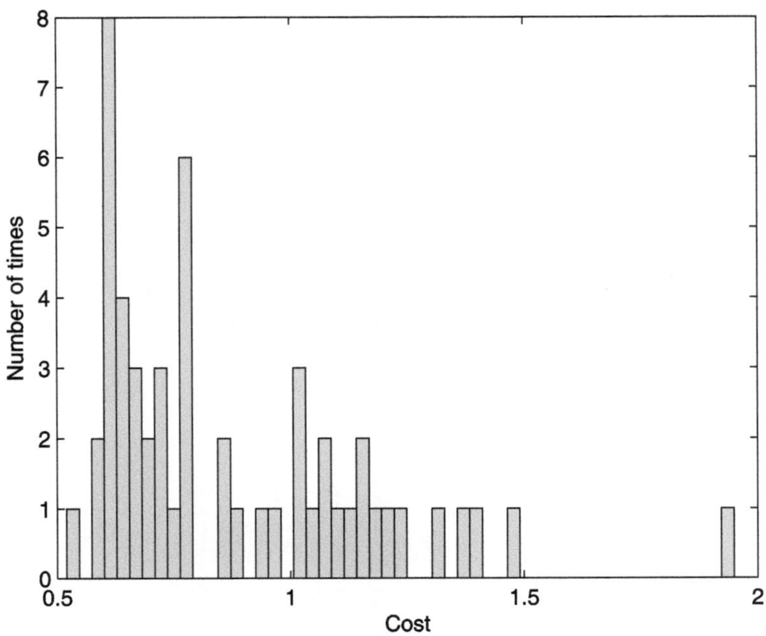

Fig. 8.40 Histogram plot obtained for a $T_d(z)$ with $a = 0.90$ using IV

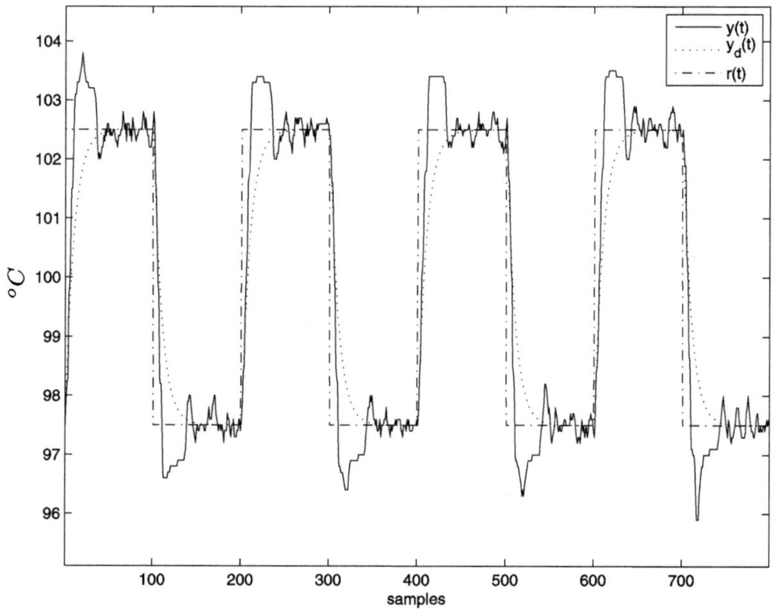

Fig. 8.41 Output signal with the initial controller

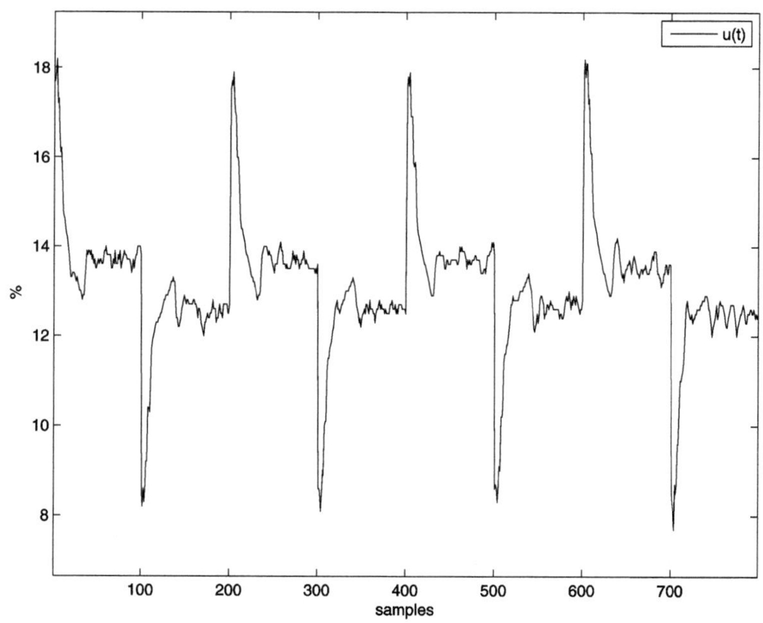

Fig. 8.42 Input signal with the initial controller

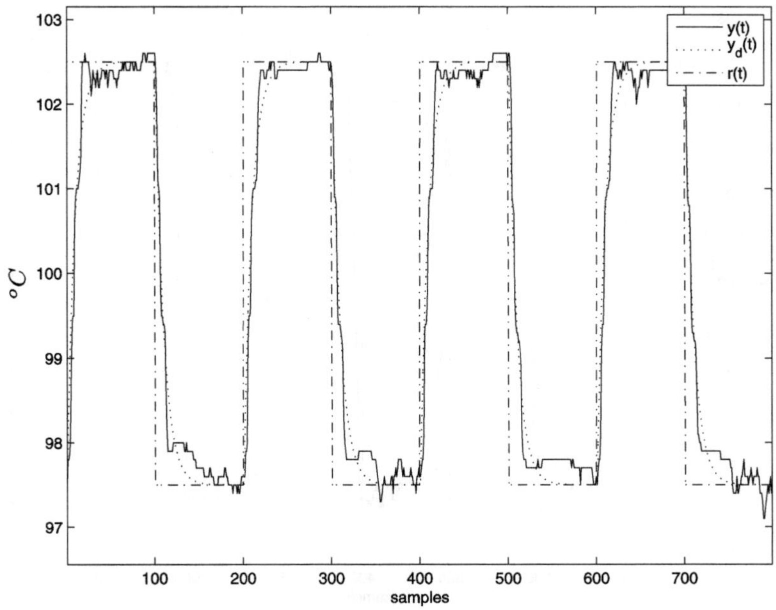

Fig. 8.43 Output signal with the controller obtained at iteration nine of the IFT method

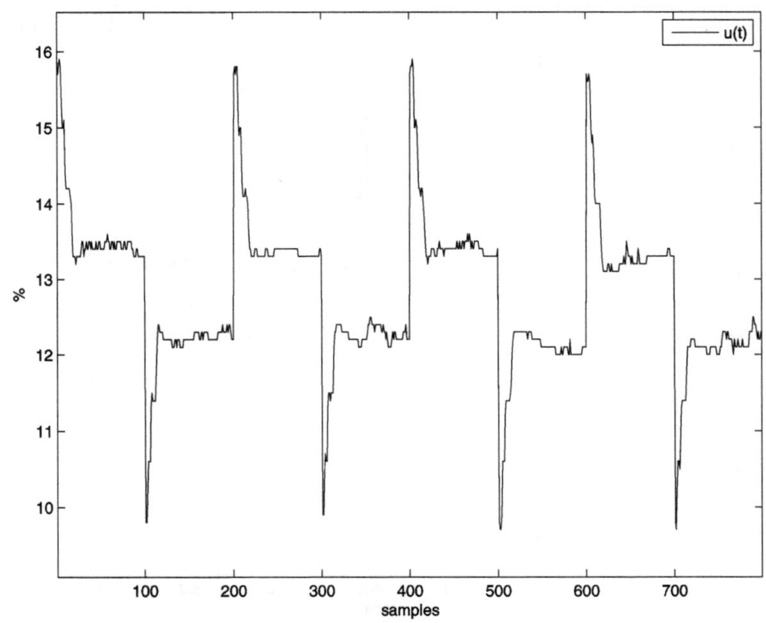

Fig. 8.44 Controller's output with the controller obtained at iteration nine of the IFT method

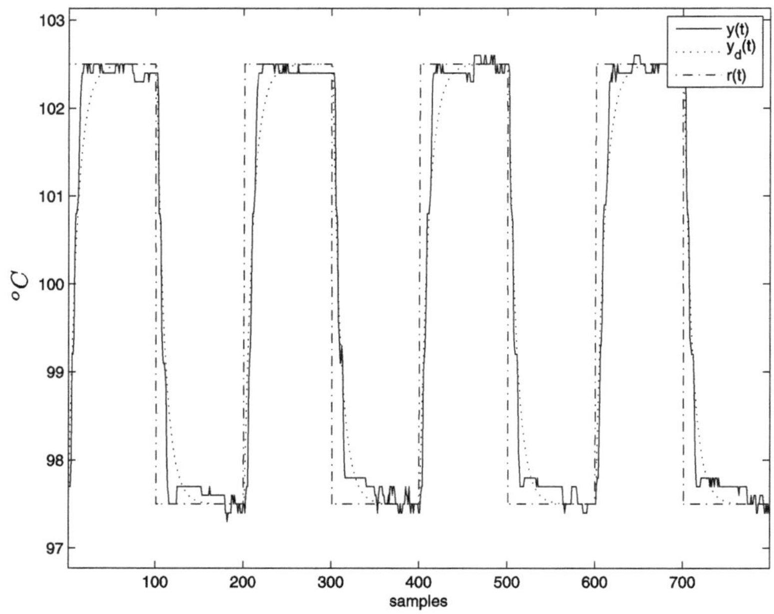

Fig. 8.45 Output signal with the controller obtained at iteration nine of the CbT method

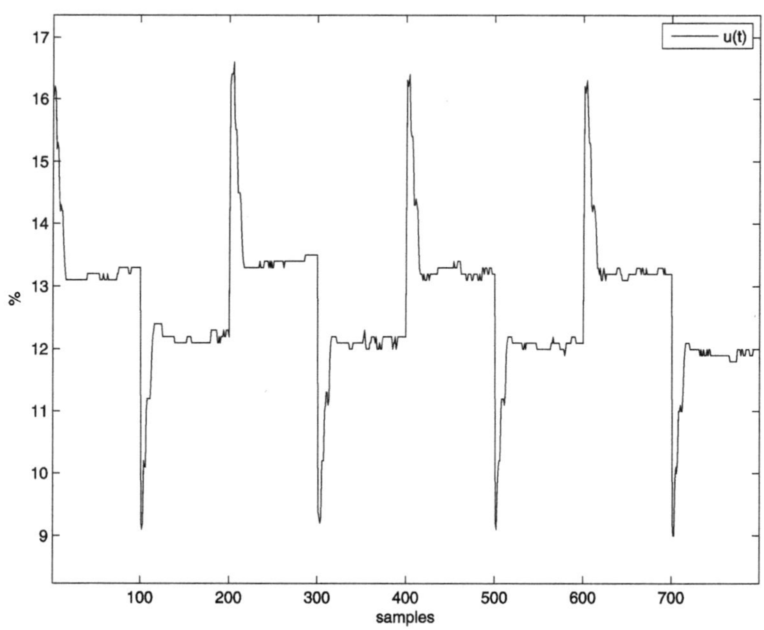

Fig. 8.46 Controller's output with the controller obtained at iteration nine of the CbT method

Table 8.9 Controller evolution using IFT

i	ρ_i	$\check{J}_T(\rho_i)$
0	[1.0000 −0.9000]	0.7314983047
1	[0.9127 −0.9029]	0.1357317944
2	[0.9211 −0.9026]	0.1053474675
3	[0.8083 −0.7898]	0.0771399942
4	[0.7363 −0.7178]	0.0693600716
5	[0.7046 −0.6861]	0.0661818423
6	[0.6987 −0.6801]	0.0645949295
7	[0.7252 −0.7057]	0.0716331607
8	[0.6988 −0.6796]	0.0655598452
9	[0.7047 −0.6854]	0.0658755918

Table 8.10 Controller evolution using CbT

i	ρ_i	$\check{J}_T(\rho_i)$
0	[1.0000 −0.9000]	0.6581802296
1	[0.9273 −0.8956]	0.1312727723
2	[0.9210 −0.8953]	0.0877604872
3	[0.9201 −0.8951]	0.1036740312
4	[0.8970 −0.8721]	0.1074603629
5	[0.8783 −0.8537]	0.1052306820
6	[0.8589 −0.8346]	0.0957999872
7	[0.8394 −0.8152]	0.0908681376
8	[0.8205 −0.7965]	0.0842272289
9	[0.8098 −0.7860]	0.0860577893

References

1. Quanser Inc., *DC Motor Control Trainer: User Manual* (2009)

MIX
Papier aus verantwortungsvollen Quellen
Paper from responsible sources
FSC® C105338

If you have any concerns about our products,
you can contact us on
ProductSafety@springernature.com

In case Publisher is established outside the EU,
the EU authorized representative is:
Springer Nature Customer Service Center GmbH
Europaplatz 3, 69115 Heidelberg, Germany

Printed by Libri Plureos GmbH
in Hamburg, Germany